Synthetic Biology: Engineering, Evolution, and Design (SEED 2021)

Online
15-18 June 2021

ISBN: 978-1-7138-4437-2

Synthetic Biology: Engineering, Evolution, and Design (SEED 2021)

Online
15-18 June 2021

AIChE
The Global Home of Chemical Engineers

Printed from e-media with permission by:

Curran Associates, Inc.
57 Morehouse Lane
Red Hook, NY 12571

Printed with permission by Curran Associates, Inc. (2022)

For permission requests, please contact AIChE
at the address below.

AIChE
120 Wall Street, FL 23
New York, NY 10005-4020

Phone: (800) 242-4363
Fax: (203) 775-5177

www.aiche.org

Additional copies of this publication are available from:

Curran Associates, Inc.
57 Morehouse Lane
Red Hook, NY 12571 USA
Phone: 845-758-0400
Fax: 845-758-2633
Email: curran@proceedings.com
Web: www.proceedings.com

TABLE OF CONTENTS

SYNTHETIC BIOLOGY FOR SAFETY, SECURITY, AND DEFENSE

KEYNOTE DAY 3

CELL-FREE SYNTHETIC BIOLOGY

COMPUTATION, ARTIFICIAL INTELLIGENCE, AND MACHINE LEARNING FOR BIOLOGICAL DESIGN

KEYNOTE DAY 4

EMERGING APPLICATIONS

EMERGING TECHNOLOGIES

POSTER SESSION

Learning to Program Cellular Machines: Engineering Cells to Treat Disease, Build Tissues, and Elucidate Principles

Lim, W. - Presenter, University of California, San Francisco

The ability to use living cells as a powerful new platform for therapeutics has the potential to transform medicine. With newfound capabilities in genetic engineering and an encyclopedic knowledge of molecular components, we have the potential to engineer cells as smart sense-and-response systems that can be tailored to address the unmet needs of complex tissue-based diseases such as cancer, autoimmunity and degeneration. Living cells, particularly immune cells, have the unique capability to infiltrate, recognize, coax and kill diseased tissues. Conversely, this cell engineering framework also presents a powerful way to test and systematically understand the general principles by which molecular components are networked to create regulatory circuits capable of precise and robust decision making. In the new science of cellular engineering, many of the most exciting challenges and roadblocks lie in developing general modules for cell sensing, recognition, communication, feedback control and decision making -- modules that could be flexibly used to rapidly and predictably program diverse cellular behaviors. I will describe our efforts to construct such modular cell engineering platforms and to use them to build smarter immune cell therapies, as well as to build self-organizing multi-cellular systems.

Interpreting the evolution of SARS-CoV-2

Bloom, J. - Presenter, California Institute of Technology

Coronaviruses undergo rapid evolution that has the potential to erode recognition by antibody immunity. I will describe the use of deep mutational scanning to generate large-scale maps of the antigenic effects of mutations. These maps are useful for interpreting the evolution of the virus to determine when mutations might alter recognition by antibody therapeutics and vaccine-induced immunity.

Programming Synthesis and Assembly of Viral Parts in 2D Compartments

Vonshak, O. - Presenter, Weizmann Institute of Science

Divon, Y., Weizmann Institute of Science

Noireaux, V., University of Minnesota

Daube, S. S., Weizmann Institute of Science

Bar-Ziv, R., Weizmann Institute of Science

The assembly of protein machines, such as viruses, in cells is precise, rapid, and coupled to protein synthesis with regulation in space and time. Reconstitution of this process outside the cell could provide means to decipher viral assembly lines and facilitate the discovery of drugs that disrupt assembly with therapeutic potential. We present quasi-2D silicon compartments that enable programming of protein assembly lines by local synthesis from surface-immobilized synthetic gene brushes[1]. Local synthesis and surface capture of complexes provide high assembly yield and sensitive detection of spatially resolved assembly intermediates. The 3D geometry of the compartment and the 2D pattern of gene brushes were found to dictate the yield and mode of assembly steps. This methodology enables deciphering, reconstructing, and designing biological machine assemblies, as well as combating pathogens.

Reference:

Vonshak, O.; Divon, Y.; FÃ¶rste, S.; Garenne, D.; Noireaux, V.; Lipowsky, R.; Rudorf, S.; Daube, S. S.; Bar-Ziv, R. H. Programming multi-protein assembly by gene-brush patterns and two-dimensional compartment geometry. *Nat. Nanotechnol.* **15,** 783Ã¢â‚¬â€œ791 (2020). https://doi.org/10.1038/s41565-020-0720-7

Synthetic Repertoires Derived from Convalescent COVID-19 Patients Enable Discovery of Sars-Cov-2 Neutralizing Antibodies and a Novel Quaternary Binding Modality

Goike, J. - Presenter, University of Texas at Austin

Gollihar, J., US Army Research Laboratory- South

The ongoing evolution of SARS-CoV-2 into more easily transmissible and infectious variants has sparked concern over the continued effectiveness of existing therapeutic antibodies and vaccines. Hence, together with increased genomic surveillance, methods to rapidly develop and assess effective interventions are critically needed. Here we report the discovery of SARS-CoV-2 neutralizing antibodies isolated from COVID-19 patients using a high-throughput platform. Antibodies were identified from unpaired donor B-cell and serum repertoires using yeast surface display, proteomics, and public light chain screening. Cryo-EM and functional characterization of the antibodies identified N3-1, an antibody that binds avidly (Kd,app = 68 pM) to the receptor binding domain (RBD) of the spike protein and robustly neutralizes the virus *in vitro*. This antibody likely binds all three RBDs of the trimeric spike protein with a single IgG. Importantly, N3-1 equivalently binds spike proteins from emerging SARS-CoV-2 variants of concern, neutralizes UK variant B.1.1.7, and binds SARS-CoV spike with nanomolar affinity. Taken together, the strategies described herein will prove broadly applicable in interrogating adaptive immunity and developing rapid response biological countermeasures to emerging pathogens.

A Viral Vector Engineered for Improved Noninvasive Gene Delivery to the Brain

Authors:

Szablowski, J. O. - Presenter, California Institute of Technology

Li, H., Caltech

Heath, J., Caltech

Shapiro, M. G., California Institute of Technology

Gene delivery to the brain with adeno-associated virus (AAV) has been a mainstay of in vivo neuroscience research and has significant therapeutic potential. Focused ultrasound (FUS) can be used to open the blood-brain barrier (FUS-BBBO) to enable noninvasive delivery of AAVs into the brain from the blood. However, the delivery of large molecules, e.g. AAVs, requires FUS pressures close to the safety margins and intravenous injection of AAVs leads to off-target peripheral tissue transduction. Here, we engineer a new AAV with improved brain transduction efficiency and tissue specificity after FUS-BBBO. To obtain a new AAV strain optimized for FUS-BBBO we modified a viral vector engineering method called Cre-recombination-based AAV targeted evolution (CREATE). In CREATE, a library of AAVs with mutated capsid proteins is injected intravenously into mice. When a particular AAV clone transduces a cell expressing Cre, its viral genome is modified and becomes detectable by a Cre-dependent PCR. We modified CREATE to engineer a FUS-BBBO-optimized AAV, which we called AAV.FUS. We first generated a viral vector library with a capsid modified by insertion of a randomized 7 amino-acid sequence between residues 588 and 589 of the AAV9 capsid. We then used FUS-BBBO to deliver the AAV library to one hemisphere of hSyn1-Cre mice, extracted the tissue, and identified 5 clones specifically enriched in the FUS-targeted regions. We observed improvement in transduction of all candidates compared to AAV9 (20-130% improvement). At the same time, each serotype transduced the liver less effectively (range: 13 â€“ 643% reduction compared to AAV9). The top AAV.FUS candidate showed a 12.1-fold improvement in overall brain-tissue specificity after FUS-BBBO, 56% improvement in neuronal tropism, and regional transduction efficiency with hippocampus showing 4.7-fold better delivery than AAV9 (32-fold improvement in liver/hippocampus transduction ratio). Overall, FUS-BBBO gene delivery was improved by AAV engineering.

Delivering RNA outside the liver by testing thousands of nanoparticles in vivo using DNA barcodes

Dahlman, J. - Presenter, Georgia Institute of Technology

DNA and RNA can manipulate the expression of any gene, making these molecules promising drugs. However, whether the drug is comprised of DNA, siRNA, mRNA, lncRNA, or another nucleic acid, it is limited by one problem: drug delivery. Chemists design thousands of distinct nanoparticles to deliver DNA or RNA to the desired cell type. However, after nanoparticles are synthesized, their ability to deliver drugs is evaluated using in vitro systems devoid of a liver, kidney, spleen, immune system, pulsatile blood flow, and other selection pressures known to affect nanoparticle delivery in vivo.
We have designed a series of increasingly advanced DNA barcoding platforms to quantify how thousands of nanoparticles deliver nucleic acids in vivo. Our goal is to quantify how up to 300 nanoparticles deliver DNA, mRNA, ASOs, or siRNA into up to 30 cell types, all in a single animal. To analyze these large in vivo drug delivery datasets, we have also developed an open source bioinformatics pipeline to iteratively â€˜evolveâ€™ nanoparticles that target cells in vivo. Using this high throughput, iterative, in vivo approach, we have identified nanoparticles with tropism to many novel cell types, in many different tissues.

Our data demonstrate that barcoded LNPs can elucidate fundamental questions about in vivo nanoparticle delivery. More generally, the data suggest that it may be feasible to perform very high throughput in vivo studies in the coming years.

Self-Replicating mRNA Therapeutics on Cancer Immunotherapy-"Multifunctional Oncolytic Nanoparticles Deliver Self-Replicating IL-12 RNA to Eliminate Established Tumors and Prime Systemic Immunity"

Li, Y. - Presenter, Synthetic Biological Center, MIT

mRNA vaccines against COVID-19 pandemic by Pfizer and Moderna saved and are saving millions of lives. The applications of mRNA therapeutics are also of great interest for cancer immunotherapy. Though therapies that synergistically stimulate immunogenic cancer cell death (ICD), inflammation and immune priming are important directions for cancer immunotherapy. However, even multi-agent therapies often fail to trigger all of the steps necessary for self-sustaining antitumor immunity. Here we describe self-replicating RNAs (replicon) encapsulated in lipid nanoparticles (LNP), which combine three key elements to engineer the tumor as â€œnon-selfâ€ and manipulate the immune responses to reject the â€œnon-selfâ€ tumor: (1) an LNP composition that potently promotes ICD, (2) RNA that stimulates danger sensors in transfected cells and (3) RNA-encoded interleukin (IL)-12 for modulation of immune cells. Intratumoral administration of LNP-replicons led to high-level expression of IL-12, stimulation of a type I interferon response and cancer cell ICD, resulting in a highly inflamed tumor microenvironment and priming of systemic antitumor immunity. In several mouse models of cancer, a single intratumoral injection of LNP-replicons eradicated large established tumors, induced protective immune memory and enabled regression of distal uninjected tumors. LNP-replicons, a kind of mRNA therapeutics, are thus a promising multifunctional single-agent immunotherapeutic.

Synthetic Biology in or near the Clinic

Collins, J. J. - Presenter, Massachusetts Institute of Technology

Synthetic biology is bringing together engineers and biologists to model, design and construct biological circuits out of proteins, genes and other bits of DNA, and to use these circuits to rewire and reprogram living organisms and cell-free systems. Here we highlight synthetic biology advances that have led to novel therapeutics, diagnostics and vaccines.

Breakthroughs in Plant Based PHB Production

Authors:

Snell, K. D. - Presenter, Yield10 Bioscience

Malik, M. R., Metabolix Oilseeds

Peoples, O. P., Yield10 Bioscience

Crops have the potential to not only be a source of food and fuel but also a source of renewable materials. One of the most interesting classes of materials are the polyhydroxyalkanoate (PHA) family of microbial carbon and energy storage polymers. PHAs are targeted as replacements for petroleum derived plastics but can also be used for water treatment and enhanced feeds. Oilseeds are an ideal production platform for PHAs since seed metabolism is well suited to the polymer biosynthetic pathways and multiple co-products can be harvested (polymer, oil, meal) increasing the value of the seed.

Yield10 and its predecessor company, Metabolix, have a long history of using synthetic biology to develop advanced PHA production organisms including microbes and plants. This capability includes developing novel biosynthetic pathways to multiple co-polymers from cellular metabolites enabling the production of PHAs with a wide range of properties in bacteria. Recently Yield10 has made significant progress producing polyhydroxybutyrate (PHB), the simplest member of the PHA family, in the oilseed *Camelina sativa.* We previously demonstrated production of high levels of PHB in Camelina seeds by targeting enzymes for PHB synthesis to the seed plastids, however seedlings were severely impaired in emergence and survival. Yield10 has demonstrated a cytosolic PHB production pathway with an ER targeted polymerizing enzyme PHA synthase that produced PHB levels up to 10.2% of the seed weight with good seedling survival. These lines have been tested in small scale field trials and seed from the best lines is currently being scaled up in the field. These results are a significant step forward towards commercial production of PHB in plants, enabling a direct link between low cost large scale agricultural production of this exciting biodegradable plastic and the packaging and foodservice segment of the 350 million ton per year global plastics market.

Universal Growth Switch for Two-Stage Bioprocess in Bacteria

Authors:

Joers, A. - Presenter, University of Tartu

Kasari, M., University of Tartu

Kasari, V., University of Tartu

Karmas, M., University of Tartu

Fast-growing bacteria spare little resources for product production whereas highly productive strains grow slowly. This growth - production tradeoff limits the productivity of many processes in industrial biotechnology. Two-stage bioprocess is a workaround to this dilemma - at first, cells are grown with maximum speed without significant product production, then the growth is switched off and all available resources are channelled to product synthesis. Switching off growth while leaving cellular metabolism active is not an easy task and existing solutions are usually specific to each product. Here we describe a novel universal growth switch of bacteria that allows to stop their growth at any given culture density. The switch is carried out by removing oriC from E. coli genome using site-specific recombinases. This prevents any further initiation of genome replication, as a result cells stop their growth, but retain active metabolism. We demonstrate that in the switched cells the expression of protein of interest is increased several times and also the period of metabolic activity is significantly longer as compared to control cells. This technology has the potential to increase the productivity of many processes in industrial biotechnology and the prospect to make production of several products economically feasible.

Zymergen's Biofacturing Platform

Kimball, A. - Presenter, Zymergen

At Zymergen, we partner with Nature to make novel, breakthrough products and materials. These products rely on the molecular diversity found in Nature to provide functionality unavailable from conventional sources. Given a market need, we can quickly develop and manufacture a relevant product. Integral to our ability to accomplish this is our Biofacturing Platform: an end-to-end integrated system for product discovery, microbe creation, and scale-up.

This talk will introduce Zymergen's Biofacturing platform and highlight some key technologies that enable identification of a relevant application-specific active ingredient molecule and rapid engineering of a prototype microbe that produces it.

Developing an International Common Mechanism for DNA Synthesis Screening

Authors:

Wheeler, N. - Presenter, University of Birmingham

Advances in biotechnology such as cheaper DNA synthesis and widespread access to gene-editing tools have given a wider segment of the population the ability to manipulate biological agents and systems. Synthetic biology makes it possible to synthesize genomes and generate viable copies of naturally occurring organisms in the laboratory or modify easily accessible organisms to enhance their pathogenicity. A vital tool for performing this work is DNA synthesis. With the market poised to transition from chemical to enzymatic DNA synthesis with higher throughput and an increasing focus on research regarding the genomics of pandemic threats, now is the time to secure international agreement on the use of DNA screening methods capable of detecting historical and emerging threats.

Currently, no governments require screening of DNA orders, and developing, implementing, and maintaining screening procedures is becoming increasingly expensive relative to other business costs, creating an economic disincentive for companies to do this. DNA synthesis providers have expressed a desire for shared assurance of reliable screening across the industry. Satisfying this demand requires an international, cost-effective, and sustainable common mechanism to prevent illicit DNA synthesis.

This talk will outline a mechanism developed in consultation with a technical consortium comprising major representatives from the regulatory, industry, and community biology sectors for screening synthetic DNA orders. The mechanism allows companies to quickly ascertain relevant information for proceeding with the order or flagging it for further attention. We welcome feedback on the proposed mechanism and encourage participation from the community in its continued development.

Discovery at Depth: A Whole Cell Bioreporter for Subterranean Detection of Chemical Agents

Authors:

LeSassier, D. S. - Presenter, Signature Science

Keplinger, N. C., Signature Science, LLC

Salazar, A. J., Texas A&M University

Ludolph, B. C., Signature Science, LLC

Kappell, A. D., Signature Science, LLC

Van Hoorn, H. E., Signature Science, LLC

Tashner, B. E., Signature Science, LLC

Tweddell, W. M., Signature Science, LLC

Straight, S. D., Signature Science, LLC

Herman, J. K., Texas A&M University

Hewitt, F. C., Signature Science, LLC

Despite advances in synthetic biology, a robust bioreporter system for subterranean detection (â‰¥ 1 meter) of buried weapons signatures or natural resources has not been produced or deployed at scale in a soil-based matrix. To meet these challenges, we developed the Spore Amplified Detection (SPADe) system as part of a DARPA-funded study. This living reporter system utilizes engineered *Bacillus subtilis*, a native soil bacterium capable of sporulating, in combination with modular chemical agent sensing of subsurface signatures to produce volatile organic compounds (VOCs) that are compatible with standoff detection. Key aspects were characterized, including chassis dissemination in soil up to a meter deep, chemical signature detection and VOC production in a variety of matrices, and standoff detection by hyperspectral imaging. Utilizing vegetative and spore forms, we demonstrated chassis dissemination through a soil column at 0.5- and 1.0-meter depths. IPTG was initially leveraged as a chemical surrogate, due to the robust nature of induction. The engineered chassis produced VOCs ethylene and methyl bromide in response to IPTG in liquid culture, detected by GC/MS, and within a soil profile, producing detectable ethylene out to 10 days. Focusing on more realistic chemical weapon signatures, the chassis was engineered for reactive chlorine species detection. Similar testing in liquid medium and soil was performed, with successful signature detection and ethylene production. To validate standoff detection compatibility, bacterially produced ethylene, from both liquid and soil matrices, was collected in a meter-long chamber and measured at a 10-meter distance using the Telops Hyper-Cam LW hyperspectral camera. Ethylene was detectable from both scenarios, down to ~15 ppmv. This works supports SPADe as a modular system for detection of chemical targets within complicated environments and highlights the potential for this system to be tailored to a range of soil-based and subterranean applications, such as for defense, remediation, or prospecting.

Distributed Information Encoding Using Self-Organized Patterns

Lu, J. - Presenter, Duke University

Tsoi, R., Duke University

You, L., Department of Molecular Genetics and Microbiology

Luo, N., Duke University

Wang, S., Duke University

Ha, Y., Duke

Dynamical systems often generate distinct outputs according to different initial conditions, and one could infer the corresponding input configuration given an output. This property captures the essence of information encoding and decoding. Past studies have explored the possibility of encoding using chaos, where small changes in initial condition can lead to drastically different outputs. While the high sensitivity makes malicious attacks without prior knowledge of the system difficult, the unavoidable noise could make decoding challenging for the designated recipient. In contrast, biological self-organized patterns share global similarity but vary in detail due to random noise under the same or similar conditions, which could balance the trade-off between reliable decoding and security. Here, we demonstrated the use of self-organized bacterial colony patterns, combined with machine learning (ML), to achieve distributed information encoding and decoding with guaranteed security.

To encode, a message is converted into cell seeding configuration followed by colony growth, during which a colony pattern develops; to decode, we input the pattern into a trained CNN to convert it back to the original message. We propose two criteria for choosing suitable patterning systems: sufficient visual complexity and weakly chaotic mapping. Both are determined by the patterning dynamic and encoding setup. By modulating these factors, we could tune the trade-off among encoding capacity, decoding accuracy and security, characterized by ML decoding performance. We also implemented ensemble techniques for enhancing decoding reliability and making full use of the expensive-to-obtain patterning data. Moreover, our approach was combined with established cryptography techniques (e.g., encryption and hashing) to further enhance the security.

Deep learning applied to Genomics: Deep semantic DNA and protein representation

Richardson, T. - Presenter, Denovium Inc

We are witnessing an explosion in biological sequence information stored in databases such as GenBank and UniProt, largely driven by massive improvements in sequencing technology and resulting in the availability of tens of thousands of genomes and metagenomes. The utility of each genome depends on high quality annotation. However, assignment of function to proteins with no known homologs is still an unsolved problem. I will discuss how Denovium developed a state-of-the-art artificial intelligence (AI) platform, the Denovium Engine, which is capable of finding and assigning function to genes directly from assembled DNA. Our DL models have been developed to be capable of interpreting proteins from primary amino acid sequence and learning multiple protein features implicitly in a single step. Denoviumâ€™s DL protein model encodes proteins in high-dimensional representations (embeddings) allowing the accurate assignment of over 700,000 labels for 25 distinct tasks. This model can be utilized to rapidly search over sequence databases to find proteins that have <25% amino acid identity to known enzymes. I will also discuss how these models can be leveraged to evolve proteins and looking into the future how we may be able to generate proteins de novo. Denovium was recently acquired by Absci and is currently being applied to the design of therapeutic proteins and associated producing cell lines.

A Partially Self-Regenerating Synthetic Cell

Lavickova, B. - Presenter, École polytechnique fédérale de Lausanne

Maerkl, S. J., Ecole Polytechnique Federale de Lausanne

Laohakunakorn, N., Ecole Polytechnique Federale de Lausanne

Self-regeneration is a fundamental function of all living systems. Here we demonstrate partial molecular self-regeneration in a synthetic cell. By implementing a minimal transcription- translation system within microfluidic reactors, the system is able to regenerate essential protein components from DNA templates and sustain synthesis activity for over a day. By quantitating genotype-phenotype relationships combined with computational modeling we find that minimizing resource competition and optimizing resource allocation are both criti- cally important for achieving robust system function. With this understanding, we achieve simultaneous regeneration of multiple proteins by determining the required DNA ratios necessary for sustained self-regeneration. This work introduces a conceptual and experi- mental framework for the development of a self-replicating synthetic cell.

Cell-Free Biosensing for Environmental and Performance Monitoring

Beabout, K. - Presenter, UES, Inc.

Breedon, A. M. E., UES, Inc.

Goodson, M., Air Force Research Laboratory

Harbaugh, S., Air Force Research Laboratory

Chávez, J., Air Force Research Laboratory

Cell-free expression systems are an exciting platform for the detection of environmental hazards and human performance biomarkers. Once lyophilized onto paper, cell-free biosensors can be transported to austere locations and rehydrated onsite for field sensing applications. Here, we show the development of a cell-free sensor that detects bile acids in both environmental samples and human biofluids. Secondary bile acids, including deoxycholic acid (DCA) and lithocholic acid (LCA), are indicators of fecal pollution in water and biomarkers for certain disease states. Importantly, a cell-free bile acid sensor could potentially detect fecal contamination in water in less than two hours, an improvement over traditional approaches with fecal coliform cultures that require greater than 24 hours of incubation. Our bile acid sensor uses BreR, a TetR-like repressor from *Vibrio cholorae*, and was successfully transferred from whole cells to cell-free with an approximately ten-fold increase in DCA sensitivity. Further characterization of the sensor in cell-free showed a strong response to several secondary bile acids. Therefore, we next explored the ability of this sensor to detect bile acids in human feces and wastewater samples. To further broaden the application space of our sensor, we also established strategies to mitigate the inhibitory effects of human blood serum on the functionality of the sensor for biomarker detection in blood. Ongoing work includes optimization of the sensor in a lyophilized paper-based assay. Additionally, we have shown that several biosensors can be transferred from whole cells to cell-free systems with an improvement in analyte sensitivity, this seems to be a common feature we are currently exploring. In summary, our bile acid sensor highlights the utility of cell-free systems for rapid onsite sensing of biomarkers and analytes for a variety of different applications.

The CRISPR Diagnostic platform: Detection of SARS-CoV-2

Fasching, C. - Presenter, Mammoth Biosciences

New CRISPR systems targeting DNA or RNA have been shown to exhibit non-specific trans-nuclease activities when bound to their respective targets. Mammoth Biosciences and others are leveraging these activities to create simple rapid sensitive and inexpensive diagnostic technologies. In response to the emergence of the second severe acute respiratory syndrome betacoronavirus (SARS-CoV-2) late in 2019, we turned our versatile technology towards developing a COVID-19 diagnostic assay. Initially, we established a reagent set using our original assay design, which we implemented in a clinical laboratory. We have subsequently expanded our technology in the development of DETECTRBOOSTTM, a high through-put CRISPR-based platform for SARS-CoV-2 testing in a centralized location.

Guiding synthetic biology via Automated Recommendation Tool (ART)

Radivojevic, T. - Presenter, Joint BioEnergy Institute

One of the most important challenges in bioengineering is effectively using -omics data to guide metabolic engineering towards higher production levels. Here, we present the Automated Recommendation Tool (ART), a tool that leverages machine learning and probabilistic modeling techniques to guide synthetic biology in a systematic fashion, without the need for a full mechanistic understanding of the biological system. ART provides a set of recommendations for the next engineering cycle, alongside probabilistic predictions of their outcomes. It can be used as a python library or through a web-based graphical frontend that does not require coding expertise.

Machine Learning Guided Workflow for Ribosome Binding Site Engineering

Holowko, M. B. - Presenter, CSIRO

Ong, C. S., CSIRO

Hayman Zumpe, H., CSIRO

Zhang, M., Australian National University

Fine control of gene expression is extremely important for engineered biological systems. Such control can be achieved through engineering transcriptional and translation control elements, including the Ribosome Binding Site (RBS). Unfortunately, our understanding of how RBSs perform *in vivo* is insufficiently well understood to allow predictable, reliable design of RBSs at the level of finesse needed for some applications. To address this problem, we have created a machine learning (ML) enabled workflow for the design of bacterial RBSs. We used Gaussian Process Regression for prediction of relative transcription initiation rates and the Upper Confidence Bound-based Bandit algorithm for recommendation of genetic designs to be tested *in vitro*. The Bandit algorithm addresses the exploration-exploitation trade-off, balancing the search for highly novel sequences with the desire for strong RBSs. We have integrated the ML algorithms with laboratory automation and high-throughput processes, creating a robust workflow for the design of custom RBSs. By employing these techniques, we were able to increase the reliability and reproducibility of results and increase the confidence in the design process. Using our workflow, we generated a novel library of diverse RBSs with a wide range of expression levels. Notably, a high number of these sites demonstrate translation initiation rates equalling or exceeding the currently known strong RBSs. Additionally, this work elucidated some design guidelines, including the favourable level of difference between two sequences and efficient numerical representation of the analysed DNA/RNA sequence. In summary, by employing both machine learning and high-throughput laboratory methods we have created a workflow for creation of small genetic devices enabling efficient generation of parts with required characteristics. As a next step, we hope to expand this workflow to more complicated parts including promoters and terminators.

Machine-Guided Cell-Fate Engineering

Appleton, E. - Presenter, Harvard Medical School

Church, G., Harvard Medical School

Stem cells are the progenitor cells of all differentiating multi-cellular organisms. In principle, it is possible to differentiate these cells into any other type of cell, which can then be used for many different possible therapeutic or diagnostic applications. The creation of induced pluripotent stem cells (iPSCs) has enabled scientists to explore the derivation of many types of cells. While there are many general approaches for cell-fate engineering, one of the fastest and most efficient approaches is transcription factor (TF) over-expression. Over-expression of specific combinations of TFs is often a reliable method to differentiate stem cells, but since there are at least 1732 transcription factors in the human genome, selecting the right combination to differentiate iPSCs directly into other cell-types is a difficult task. Here were describe a machine-learning (ML) pipeline, called CellCartographer, for using chromatin accessibility next-generation sequencing (NGS) data to produce a multiplex TF pooled-screen for converting stem cells into other cell types. We then describe a barcoded bulk RNA-seq method for refining the set of TFs using iterative NGS experiments. We validate this method by differentiating stem cells into six medically-relevant cell types with the human TFome originating from all germ layers: cytotoxic T-cells, regulatory T-cells, B-cells, microglia, type II astrocytes, and hepatocytes. We demonstrate iterative improvement in differentiation efficiency and functionally characterize the cell lines to validate fast, robust, and functionally accurate differentiation of stem cells into cell types useful for downstream therapeutic and diagnostic pipelines.

Engineered Autonomous Control of Metabolic Pathways

Prather, K. - Presenter, Massachusetts Institute of Technology

Microbial systems offer the opportunity to produce a wide variety of chemical compounds in a sustainable fashion. Economical production, however, requires processes that operate with high titer, productivity, and yield. One challenge towards maximizing yields is the need to use substrate for biomass, resulting in a competing pathway that cannot merely be eliminated. Productivities may also be significantly influenced by the timing of expression of genes in the production pathway. Dynamic metabolic engineering has emerged as a means to address these and other impediments in strain performance. Ideally, the triggers for dynamic control would be autonomous, that is, independent of any external intervention by the operator. We have developed such autonomous devices based on pathway-independent quorum-sensing circuits and have demonstrated their utility across several distinct metabolic pathways and with varying levels of complexity. In this talk, I will describe our approach for development of these Metabolite Valves and results to date from their implementation.

HyperXpress: High-throughput, one-pot DNA assembly and protein production in low microliter volumes using cell-free extracts

Kabisch, J. - Presenter, Technical University Darmstadt

For academic and industrial life science research, efficient DNA assembly methods and gene expression systems are an essential tool for realizing numerous applications and gaining far-reaching knowledge. In this context, high-throughput methods have a prominent position by allowing a large number of experiments in a short time.[1] Moreover, one-pot processes are favored for the actual experiments because they are quick, inexpensive and easy to use.[2]

HyperXpress is intended to provide a high-throughput, one-pot process for a successive sequence of DNA assembly and *in vitro* gene expression. As part of HyperXpress, an optimized ligase cycling reaction (LCR)[3], multiply-primed rolling circle amplification (RCA)[4] and cell extract-based *in vitro* cell-free protein synthesis (CFPS)[5] are carried out step by step in a single well of a 384-well plate.

We present some of the optimization of LCR, RCA and CFPS performed to enable protein production in 3.6 Âµl in 10 hours starting from DNA fragments as well as a software addition to the DiVA-Software[6] which aids in bridging oligo design and liquid handling. As a proof-of-concept a family shuffling of GFP-ÃŸ-strands was performed. The resulting 51 variants consisted of ÃŸ-strand deletions, ÃŸ-strand-subtstitutions as well as of heterologous ÃŸ-strand additions.

1. **References**

[1]: â€žDirect assembling methodologies for high-throughput bioscreeningâ€� (2012) by Jorge I. RodrÃ-guez-DÃ©vora, Zhi-dong Shi and Tao Xu;
Biotechnol J. 2011 Dec; 6(12): 1454â€"1465. DOI: 10.1002/biot.201100100
[2]: â€žPot economy and one-pot synthesisâ€� (2016) by Yujiro Hayashi,
Chem Sci. 2016 Feb 1; 7(2): 866â€"880. DOI: 10.1039/c5sc02913a
[3]: â€žOptimization of the experimental parameters of the ligase cycling reactionâ€� (2019) by Niels Schlichting, Felix Reinhardt, Sven Jager, Michael Schmidt and Johannes Kabisch;
Synthetic Biology, Volume 4, Issue 1, 2019, ysz020, DOI: 10.1093/synbio/ysz020
[4]: â€žRapid Amplification of Plasmid and Phage DNA Using Phi29 DNA Polymerase and Multiply-Primed Rolling Circle Amplificationâ€� (2001) by Frank B. Dean, John R. Nelson, Theresa L. Giesler and Roger S. Lasken;
Genome Res. 2001 Jun; 11(6): 1095â€"1099; DOI: 10.1101/gr.180501
[5]: â€žProtocols for Implementing an Escherichia coli Based TX-TL Cell-Free Expression System for Synthetic Biologyâ€� (2013) by Zachary Z. Sun, Clarmyra A. Hayes, Jonghyeon Shin, Filippo Caschera, Richard M. Murray and Vincent Noireaux;
J Vis Exp. 2013; (79): 50762 ; DOI: 10.3791/50762
[6]: DiVA - Design, Implementation, and Verification Automation Platform, https://public-diva.jbei.org/

Prospects to Enhance Biological and Mechanical Performance in Living Building Materials

Artier, J. - Presenter, University of Colorado Boulder

Qiu, J., Colorado University Boulder

Cook, S. M., University of Colorado Boulder

Srubar, W. V. III, University of Colorado Boulder

Hubler, M. H., University of Colorado Boulder

Cameron, J. C., University of Colorado Boulder

Living building materials (LBMs) are produced by integrating viable microorganisms into inert structural scaffolds. Our framework takes advantage of microbially induced calcium carbonate precipitation (MICP) and has potential for recycling and self-healing, features that depends on the microorganism survival under LBM dehydrated conditions. Cyanobacteria is exposed to desiccation stress under nonaquatic ambient conditions. By introducing a desiccation protectant, the small sugar trehalose, LBMs cells were able to survive in the dried state for several days, with no negative effect on mechanical properties. We genetically engineered *Synechococcus* 7002 strains to synthetize trehalose, by introducing *treZ* and *treY* genes from *Anabaena* sp. PCC 7120, a moderate desiccation tolerant aquatic organism, under regulation of two strong promoters, a native constitutive and an IPTG inducible. Our novel LBM is composed of sand, hydrogel (gelatin) for binding aggregates, and live bacteria, capable of inducing calcium carbonate precipitation. Cyanobacteria has been shown to induce precipitation of calcite, dependent on photosynthesis, with the advantage of capturing and storing CO_2. Comparison between biotic and abiotic MICP show greater structural properties, especially superior fracture energy for biotic MICP samples. Autotrophic and heterotrophic bacterial MICP pathways, through either ureolysis or photosynthesis, were successful in mechanically enhancing the LBM, including material compression strength and flexural energy. We also tailored several LBM design factors in an effort to expand the multifunctionality of our building materials. This study shows that LBMs are an exciting new class of materials with a range of features that can be tailored for adaptation to the in-service application. This project is funded by DARPA: Engineering Living Materials Program.

Terminal Deoxynucleotidyl Transferase Acts As a Signal Sensor and Writer for Biorecording and Data Storage Applications

Bhan, N. - Presenter, Northwestern University

Castinado, A., Northwestern University

Strutz, J., Northwestern University

Kalhor, R., Harvard Medical School

Glaser, J., Northwestern University

Kording, K. P., Northwestern University

Boyden, E. S., Massachusetts Institute of Technology

Church, G. M., Harvard Medical School

Tyo, K., Northwestern University

DNA-based recording tools used for data archiving and investigating biological systems have significantly advanced the field of synthetic biology and enriched the bioeconomy. Like traditional electronic recording systems, state-of-the-art DNA-based recording systems consist of multiple components: a signal sensing, a signal transforming and a signal writing device. This multi-device architecture makes their recording characteristics intrinsically dependent on speed of each of those devices. The ideal DNA-based recorder would consist of a single device that could sense, transform and write into DNA with high accuracy and information density. We introduce terminal deoxynucleotidyl transferase (TdT) as both the signal sensor and writer, for biorecording and data storage applications. TdT, a template-independent DNA polymerase, upon interacting with the signal of interest alters its dNTP incorporation preference. The resulting single stranded DNA (ssDNA) synthesized thus functions as a molecular ticker tape. We show that *in vitro*, TdT can record various physiologically relevant signals like Co^{2+}, Ca^{2+}, Zn^{2+} concentrationsand temperature changes. Further, we record a single step change in Co^{2+} signal to within two minutes over a 60-minute period and record the rise and fall of an Co^{2+} input signal to within three minutes. Finally, we demonstrate a proof-of-concept for the TdT-based recording system in mammalian cell lines by writing â€œHello World!â€� into the genomic DNA without the need of any externally synthesized oligonucleotides. We have thus expanded the repertoire of DNA-based recording techniques by developing a â€œDNA-synthesisâ€� based system, that can record temporal environmental signals into DNA.

Co-Transcriptionally Encoded RNA Strand Displacement Circuits

Schaffter, S. - Presenter, Johns Hopkins University

Strychalski, E. A., NIST

The programmable and predictable kinetic control of DNA strand displacement reactions make them a powerful tool for engineering diverse chemical information processing circuits. Chemical reaction networks based on DNA strand displacement have demonstrated complex digital computation, molecular pattern recognition, and signal cascades and amplifiers. However, current DNA strand displacement circuits are single-use and are susceptible to degradation in environments like serum or living cells. This limits their ability to detect and process nucleic acid inputs in many practical settings. Here, we present co-transcriptional RNA strand displacement (ctRSD) circuits, where DNA strand displacement components are directly mapped to RNA transcripts, allowing strand displacement circuits to be co-transcriptionally encoded. Key to ctRSD circuit design is the ability to co-transcriptionally fold kinetically trapped RNA gates, which allows all circuit components to be produced together without significant cross reaction. We demonstrate ctRSD OR, AND, and signal amplification elements, and we integrate these elements into larger signaling cascades. We test 10 unique ctRSD gates in eight different circuit architectures and find, analogous to DSD circuits, ctRSD components are modular and composable. Finally, ctRSD circuit kinetics quantitatively match the predictions of a simple model of coupled transcription and strand displacement, enabling model-driven circuit design. We envision our robust ctRSD design rules will allow the existing information processing capabilities of DNA computing to be readily adopted for new sensing and diagnostic applications and, ultimately, for continuous real-time nucleic acid computation inside living cells.

Scalable DNA Data Storage Systems

Keung, A. J. - Presenter, North Carolina State University

DNA holds significant promise as a data storage medium due to its density, longevity, and resource and energy sustainability. These advantages arise from the inherent biomolecular structure of DNA as well as, perhaps unintuitively, the spatially disordered nature of DNA mixtures. For example, disordered pools of DNA strands confer extreme density, while arraying DNA strands on fixed substrates or within microwells would abrogate almost all of this density advantage. Furthermore, even just terabyte-sized databases would require mixtures of billions of distinct DNA strands, much larger than the sequence diversity of the human genome; thus, the unique molecular and disordered nature of DNA storage presents clear challenges arising simply from the fact that many diverse strands of DNA will be in crowded proximity with severe thermodynamically-driven limitations on file access specificity and accuracy. This prompts important discussions on how data should be organized, accessed, and manipulated. Here we argue that the novel architecture of DNA storage may actually confer novel functionalities beyond what is possible with conventional storage media, and that inspiration from natural biological processes and biochemistry can be leveraged to engineer novel and scalable functionalities into DNA databases. We will describe three examples of this approach. We will show how hierarchical file systems can be engineered to expand the number of possible unique DNA file addresses by five orders of magnitude and that individual KB sized files can be specifically accessed from large TB sized background databases. Next, we took inspiration from the processes of gene expression and strand displacements to engineer a dynamic and reusable DNA storage system that can rename, lock, and delete files while also improving encoding efficiency and information density over current random access methods. Finally, we describe how thermodynamics can be leveraged to tune biomolecular interactions and implement functions such as File Preview.

"Multi-Agent" Screening Improves the Efficiency of Directed Enzyme Evolution

- Conference: Synthetic Biology Engineering Evolution Design SEED
- Year: 2021
- Proceeding: 2021 Synthetic Biology: Engineering, Evolution & Design (SEED)
- Group: Poster Session
- Session:

Poster Presenters - Accepted

Authors:

Yang, T. - Presenter, Duke University

Lynch, M., Duke University

Ye, Z., Duke University

Enzyme evolution has enabled numerous advances in biotechnology. However, directed evolution programs can still require many iterative rounds of screening to identify optimal mutant sequences. This is due to the sparsity of the fitness landscape, which in turn, is due to "hidden" mutations that only offer improvements synergistically in combination with other mutations. These â€œhiddenâ€� mutations are only identified by evaluating mutant combinations, necessitating large combinatorial libraries or iterative rounds of screening. Here, we report a multi-agent directed evolution approach that incorporates diverse substrate analogues in the screening process. With multiple substrates acting like multiple agents navigating the fitness landscape, we are able to identify â€œhiddenâ€� mutant residues that impact substrate specificity without a need for testing numerous combinations. We initially validate this approach in engineering a malonyl-CoA synthetase for improved activity with a wide variety of non-natural substrates. We found that â€œhiddenâ€� mutations are often distant from the active site, making them hard to predict using popular structure-based methods. Interestingly, the many of the â€œhiddenâ€� mutations identified in this case are expected to destabilize interactions between elements of secondary structure, potentially affecting protein flexibility. This approach may be widely applicable to accelerate enzyme engineering. Lastly, multi-agent system inspired approaches may be more broadly useful in tackling other complex combinatorial search problems in biology.

A New Form of Bio-Active Materials: Blending Dry Cell-Free Systems and Synthetic Polymers

Lee, M. S. - Presenter, US Army CCDC CBC

Lux, M., CCDC Chemical Biological Center

Gupta, M., Air Force Research Laboratory

Kuhn, D., US Army CCDC CBC

Raig, R., UES Inc, AFRL

Peterson, G. W., Edgewood Chemical Biological Center

Biondo, J., US Army DEVCOM Chemical Biological Center

Cell-free systems have growing importance as a way to use synthetic biology tools in resource-poor environments. Lysates may be dried for storage, delivering biochemical activity for sensing or producing molecules on-demand upon rehydration at the point of need. Up to now, cell-free protein synthesis (CFPS) reactions have been studied as aqueous solutions in test tubes or absorbed onto paper or cloth. Embedding biological functionality into broadly-used materials, such as plastic polymers, has long been an attractive goal. Unfortunately, this goal has for the most part remained out of reach, often due to the fragility of biological systems outside of aqueous environments. In this work, we describe and utilize a surprising and useful feature of lyophilized cell-free lysate systems: tolerance to casting methods involving exposure to organic solvents or heat up to 90Â°C. To explore this recently discovered feature, a variety of solvents were tested and CFPS reaction components were screened for protective properties. CFPS is compatible with several polymer types and casting conditions. Tolerance to polymer casting enables the delivery of dry cell-free reactions in the form of coatings or fibers, among other processing possibilities.

A Novel DNA Purification Technology Allows for Efficient and Faster Assembly of Expression Clones for CRISPR/Cas9-Mediated Genome Editing

- Conference: Synthetic Biology Engineering Evolution Design SEED
- Year: 2021
- Proceeding: 2021 Synthetic Biology: Engineering, Evolution & Design (SEED)
- Group: Poster Session
- Session:

Poster Presenters - Accepted

Authors:

Doh, J. - Presenter, Beckman Coulter Life Sciences

The CRISPR/Cas9 technology allows for rapid and efficient genomic modifications in a variety of cellular contexts. However, the construction of expression clones can be monotonous, requiring multiple rounds of cloning and/or PCR. Due to the necessity of replicates and repeat experiments to obtain even a single desired construct, multiple rounds and hours of nucleic acid purifications are performed from post-enzymatic reactions and bacteria. Here, we describe the assembly of gRNA expression cassettes into vectors via Golden Gate and Gibson cloning technologies using two distinct DNA purification technologies in parallel. We demonstrate that the utility of a novel PCR and plasmid purification method provides significant time-savings, increased performance, and ease of use compared to traditional spin-columns. Furthermore, the efficiency of this novel system was verified by Next Generation Sequencing. The system reported here provides a much more efficient and faster process to construct expression clones for CRISPR/Cas9-mediated genome editing.

A Single-Layer Artificial Neural Network Type Architecture with Molecular Engineered Bacteria for Complex Conventional Computing

Sarkar, K. - Presenter, Saha Institute of Nuclear Physics (HBNI), Kolkata, India

Bonnerjee, D., Homi Bhabha National Institute

Srivastava, R., Saha Institute of Nuclear Physics (HBNI)

Bagh, S., Saha Institute of Nuclear Physics

Artificial neural networks (ANNs) are inspired by the human brain and they are designed to compute complex problems bearing resemblance to the way in which the human brain does. Hardware implementation of artificial neural networks has been carried out in photonics, material-based neuromorphic chips and DNA strand displacement processes. However, it has never been physically realized in living cells. Here, we adapt the basic concept of ANN and develop the single-layer ANN-type architectures with engineered bacteria to perform complex computations. We have built and optimized molecular devices within the bacterial cells where they act like artificial neuro-synapses and the bacterium carrying the molecular device behave like an artificial neuron (Bactoneuron). The externally supplied input chemicals, associated with each bactoneuron get linearly combined along with their weights and the bias into the summation function which further gets processed through a non-linear activation function to give rise to the output in terms of the fluorescent protein production. We further illustrate a general framework to build bactoneuron-based single layer ANN-type architecture from the truth table of a given function without considering its electronic circuit design where each bactoneuron is employed to perform a specific function which is the subset of the final function and at the network level, their collective behavior gives rise to the entire final function in a multi-population culture. We have also developed the sets of rules to distribute a functional truth table into the smaller parts and design the bactoneurons out of them, adjusted their weights and biases manually by engineering the molecular interactions within the bacteria and finally observed their performances at the network-level. In this way, we have developed bacteria-based single-layer ANN type architectures working as a 2-to-4 decoder and a 4-to-2 priority encoder. This work may possess a new design approach for complex biocomputing.

A Synthetic Chromatin Reader-Effector Activates Key Silenced Tumor Suppressor Genes in Triple Negative Breast Cancer Cells

Williams, N. L. - Presenter, Emory University

Shields, C., Emory University

Hong, L., Georgia Institute of Technology

Haynes, K., Emory University

Epigenetic control has arisen as a driver and druggable vulnerability in many hard-to-treat cancers. The development of epigenetic therapy is advancing as a way to reactivate tumor suppressor genes (TSGs) as an alternative to traditional chemotherapies. Triple negative breast cancer (TNBC) often shows relapse after chemotherapy and is untreatable with hormone therapy, but has shown modest responses to leading epigenetic drugs *in vitro, in vivo*, and in clinical trials. Currently, the design of epigenetic treatments for TNBC is limited to small molecule inhibitors that bind chromatin-modifying enzymes. Critical limitations such as passive gene regulation and off-target interactions outside of chromatin prevent epigenetic therapy from meeting its full potential. Our overall goal is to engineer synthetic reader-effectors to override aberrant gene expression profiles in cancer. In TNBC, CBX reader-repressor proteins are overexpressed and silence the expression of TSGs. In this study, we compare the anti-cancer effect of a synthetic reader-activator, Polycomb-based Transcription factor (PcTF), versus small molecule inhibitors. PcTF is designed to bind a biochemical signature in silenced chromatin (histone H3K27me3) and induce immediate activation of nearby genes. To support clinical translation, we identified a set of 99 TSGs that are silenced in TNBC patient samples and in model TNBC cell lines. In PcTF-expressing cells, we observed immediate activation of several of these TSGs accompanied by reduced viability in 12 - 24 hours, which suggests a faster anti-cancer effect than the epigenetic inhibitor compounds that typically require 3 - 7 days of treatment. In conclusion, we have established a strategy to compare the impact of chromatin engineering versus epigenetic drugs on gene expression and cancer cell viability *in vitro*. These results lead the way towards investigating cancer apoptosis *in vitro*, and inhibition of metastasis in a mouse model (currently underway).

A Synthetic Live Attenuated Bacterial Vaccine for Countering Autoimmune Disease

Poster Presenters - Accepted

Authors:

de Figueiredo, P. - Presenter, Texas A&M Health Science Center

Das, J., Texas A&M Health Science Center

Alaniz, R., Texas A&M Health Science Center

Song, J., Texas A&M Health Science Center

Ficht, T., Texas A&M University

Guo, F., Texas A&M Health Science

Immunotherapy has led to impressive advances in the treatment of human diseases, including autoimmune disorders; yet, its clinical outcomes remain limited by various factors, including the subversive effects of pro-inflammatory environments, and the fact that the key molecular determinants of immune tolerance remain elusive. Recently, significant interest in exploiting live attenuated bacterial strains for addressing human diseases has emerged. However, relatively few examples of the application of such vaccines for addressing rheumatoid arthritis have been reported. Here, we describe a synthetic, metabolically engineered, live attenuated vaccine strain of *Brucella melitensis* (Bm) that expresses natural compounds (indoles) that reduce tissue inflammation, polarize M2 macrophages (MÏ†), promote regulatory T cell (T_{reg}) activation, and reduce expression of pro-inflammatory cytokines. Strikingly, a combined treatment of the synthetic vaccine with an adoptive cell transfer (ACT) of regulatory T cells (T_{regs}) significantly enhanced the accumulation of M2 MÏ† and T_{reg} persistence in the inflamed tissues, resulting in the suppression of the development of autoimmune arthritis. Taken together, this work defines a novel strategy to reinforce immunotherapy against autoimmune disorders such as rheumatoid arthritis (RA), and provides new therapeutic prospects for improving immunotherapy for autoimmune disorders.

Translate this page

A Targeted Genome-Wide Approach to Elucidate and Control Bacterial Adhesion to Physicochemically Diverse Biomaterials

Call, S. - Presenter, University of Massachusetts Amherst

Ugbesia, B., University of Massachusetts Amherst

Vu, V., University of Massachusetts, Amherst.

Andrews, L. B., University of Massachusetts Amherst

Biofilm infections on indwelling biomedical devices are the leading cause of hospital infections around the world. The most common are catheter-associated infections caused by *Escherichia coli* and *Staphylococcus aureus*. Prevention and treatment of these infections typically rely on broad-spectrum antibiotics, yet these strategies often promote antibiotic resistance and are not fully effective. Elucidating the genetic elements responsible for bacterial adhesion to and biofilm formation on biomaterials could be used for the rational design of targeted antibiofouling agents that prevent infections on biomedical device surfaces. In this work, we aim to systematically map the genomes of *E. coli* and *S. aureus* to identify genetic targets that confer adhesion phenotypes and quantify their fitness using a multiplexed, high-throughput approach. In this approach, we design comprehensive genome-wide CRISPR interference (CRISPRi) gene repression libraries for *E. coli* and *S. aureus* and apply pooled selections on a platform of biomaterials (PEG and PDMS gels) that systematically vary in their physicochemical properties to uncover gene expression associated with bacterial attachment to various surfaces. We have developed and characterized CRISPRi tools in *E. coli* MG1655 using three CRISPRi systems to determine their design rules and appropriate expression levels to prevent cellular toxicity. Repression of genomic loci, including loci associated with biofilm formation, was quantified using superfolder GFP fusions. Using the determined design rules and custom Python scripts, we have designed gRNA libraries for each CRISPRi system (>40,000 designs) to target nearly every annotated gene in the MG1655 genome (95.9%â€"99.1%). For *S. aureus*, very few synthetic DNA parts have been characterized and reported in the literature. Here, we created and report what we believe to be the first characterized genetic part toolbox for *S. aureus*, including promoters, ribosome binding sites, terminators, and origins of replication. We then used the toolbox to rationally design CRISPRi tools for *S. aureus*.

Abiotic Fluorescent Probes for Rapid Screening of Membrane-Disrupting Antibiotics

Kallmyer, N. - Presenter, Iowa State University

Reuel, N., Iowa State University

Eeg, D., Iowa State University

Khor, R., Iowa State University

Roby, N., Iowa State University

As antimicrobial resistant bacteria become increasingly prevalent, the development of commercial antibiotics has counterintuitively slowed due to their costly and lengthy development timelines. Major bottlenecks in antimicrobial development are related to practical constraints with live, infectious bacteria, which require specialized (BSL 2+) facilities and limit throughput to the timescale of bacterial growth, which can span hours to days. Herein, we address these bottlenecks by developing fluorescent probes which use single-walled carbon nanotubes (SWNT) functionalized with bacteria membrane lipopolysaccharides (LPS) as abiotic proxies to Gram-negative bacteria. LPS form a â€œcoronaâ€ phase around SWNT which both solubilizes the otherwise-hydrophobic nanotube and confers surface functionality similar to that of a Gram-negative bacterium. Disruption of this corona phase is transduced by fluorescence quenching, observed in real time. Within sensor response traces, we have also observed distinct kinetic regimes which suggest different mechanisms of disruption such as binding, pore formation, and translocation. Further characterization may allow these sensors to distinguish and multiplex disruption mechanisms. Ultimately, these sensors will reduce barriers to antimicrobial discovery by enabling high-throughput testing in common labs that lack biosafety ratings.

Activating Silent Biosynthetic Gene Clusters in Streptomyces Spp. with Crispri

Ameruoso, A. - Presenter, Rice University

Chappell, J., Rice University

Antibiotic-resistant bacteria are a looming threat for healthcare, and illustrate the need for the discovery of novel antimicrobial agents. The bacterial genus *Streptomyces* represents a promising resource, as it is the most prominent producer of antimicrobial natural products. This ability lies in the expression of enzymatic pathways encoded by genomic regions called Biosynthetic Gene Clusters (BGCs). BGCs are rarely expressed under laboratory culturing conditions (*i.e.* are silent), often due to the repressive action of gene regulatory networks. Therefore, many BGCs are uncharacterized and a potential source of new compounds. Here we describe a novel approach to activate silent BGCs by using CRISPR interference (CRISPRi) transcriptional repressors to relieve repressive regulation and activate natural product synthesis. To do this, we develop a CRISPRi system for *Streptomyces* and optimize its repression capability by tuning the expression levels of its components. We then harness CRISPRi to inhibit the activity of known BGC repressors and successfully induce production of natural products. Together, this work provides a new approach to activate BGC expression in *Streptomyces*. This strategy circumvents the need for time-consuming genome engineering efforts and only requires straightforward transformation of easily designable plasmids, thus providing a simple method for BGC activation.

Advancement of Cofactor F420 Accessibility for Biotechnology By Synthetic Biology and Metabolic Engineering

Nazem-Bokaee, H. - Presenter, CSIRO

Scott, C. - Presenter, CSIRO

Shah, M., CSIRO

Anthony, J., ANU

Bashiri, G., The University of Auckland

Holowko, M. B., CSIRO

Jackson, C., Australia National University

Baker, E. N., The University of Auckland

Greening, C., Monash University

Ney, B., Australian National University

Kang, S. W., Australian National University

The deazaflavin cofactor ferredoxin 420 (F_{420}), synthesised by some Archaea and Eubacteria, is involved in energy metabolism, xenobiotic biodegradation, and biosynthesis of secondary metabolites. Due to its low redox potential, F_{420} could potentially be used in novel industrial biocatalysis. However, one major barrier for such application is the production of cofactor F_{420} at scale and at low cost. Here, through applying synthetic biology and metabolic engineering strategies, we present: (1) the development of an engineered *E. coli* production system for the biosynthesis of F_{420}, (2) improvement of F_{420} space-time production yield by the recombinant *E. coli* up to ~40-fold higher than the yield from the wildly-used F_{420} producer, *Mycobacterium smegmatis*, and (3) identification and construction of key gene knock-out *E. coli* mutants which further enhances F_{420} productivity. The bacterial F_{420} biosynthesis pathway was revised and chosen for heterologous expression in *E. coli*. The biosynthesis of main precursor for F_{420}, dehydro-F_{420}, was found to be dependent on the utilisation of phosphoenolpyruvate (PEP) in this revised pathway. In subsequent works, we showed that, in addition to the choice of carbon source as substrate, PEP is a key metabolite influencing intracellular F_{420} concentrations. We further improved F_{420} productivity by over-expressing PEP synthase. Additionally, we developed an updated genome-scale metabolic model of *E. coli* capable of predicting metabolic-level cellular response to the introduction of orthogonal F_{420} biosynthesis pathway under given growth conditions. We identified gene knock-outs resulted in improved F_{420} yields *in silico*. These results provide important insights on establishing *E. coli* as an industrial F_{420}-production system for applications in biocatalysis.

An Algorithmic Approach to Program Probiotic Bacterial Strains for In Vivo Diagnostics

Lebovich, M. - Presenter, University of Massachusetts Amherst

Andrews, L. B., University of Massachusetts Amherst

Through the implementation of designable genetic circuits, probiotic strains of bacteria can be used as non-invasive diagnostic tools for the gastrointestinal tract. For these programmed cells to report biomarkers after exiting the gut, the genetic circuits must be able to detect and record the relevant signals within the gut environment using genetically encoded memory. Complex circuits with memory would allow for multiplexed detection and reporting of biomarkers *in vivo*. We previously developed a computational approach for the scalable design of genetic circuits that contain memory, known as sequential logic circuits. However, the ways that performance of the genetic circuits and components varies for probiotic bacterial strains and growth conditions remained poorly understood. Here, we demonstrate that the theory-based approach to design sequential logic circuits from simple logic gate responses can be implemented in gut bacteria, and these circuits can be integrated with gut metabolite biosensors. For this work, we used repressors of the TetR family to construct logic NOT/NOR gates that can be composed into complex genetic circuits and those containing memory. The genetic logic gates were characterized in the probiotic strain *E. coli* Nissle 1917. Using this data, we designed and predicted the behavior of larger circuit designs. We present a set of genetic circuits that encode sequential logic and show that the circuit outputs are in close agreement with our quantitative predictions from the design algorithm. In addition, we have generated and characterized a set of genetically encoded biosensors in *E. coli* Nissle 1917 for biochemicals that may be found in the human gut, including propionate and choline. Using the genetic circuit design algorithm, these sensors were integrated into genetic circuits to sense and record specified concentration ranges of these molecules as a test case for the use of this platform for the design of diagnostic gut bacteria.

Anaerobic Conditioning of E. coli cell Lysate for Enhanced in Vitro protein Synthesis

Tamiev, D. - Presenter, Iowa State University

Dopp, J., Iowa State University

Reuel, N., Iowa State University

Cell free protein expression (CFPS) from *E. coli* cell lysate is an established chemical biology technique especially suited for synthetic biology prototyping. Common efforts to improve synthesis capacity, such as strain engineering and lysate preparation process improvements, have overlooked the opportunity to increase productivity by reducing the dependence on limited, dissolved oxygen. Here we demonstrate conditioning *E. coli* cells for anaerobic respiration prior to standard lysate preparation which increases the initial protein expression rate by four-fold and titer by 50% as compared to a panel of aerobically cultured cell lysates (custom and commercial) when using sfGFP as a reporter protein in CFPS reactions run at atmospheric conditions. In this work we demonstrate that the advantage of using anaerobically conditioned lysate is even more pronounced when CFPS reactions are conducted in oxygen-depleted environments that mimic larger vessel CFPS experiments or oxygen-limited microdroplet reactors. For the first time, we demonstrate the possibility of increasing CFPS productivity through conditioning cells to use alternative terminal acceptors (TEA) such as nitrate. Further improvements in preconditioning cell lysate are possible through optimizing expression levels of anaerobic metabolism enzymes and screening other TEA candidates.

Application of Natural RNA Regulators to Tune Biosynthesis of Nanoparticles for Enhanced Material and Functional Properties

Chen, A. - Presenter, University of Texas at Austin

Hernandez-Vargas, J., UNIVERSIDAD MICHOACANA DE SAN NICOLAS DE HIDALGO

Han, R., The University of Texas at Austin

Cortazar-Martinez, O., Cinvestav Queretaro

Gonzalez, N., University of Texas at Austin

Patel, S., The University of Texas at Austin

Keitz, B. K., The University of Texas at Austin

Luna-Barcenas, G., CINVESTAV-Querétaro

Contreras, L. M., The University of Texas at Austin

The biosynthesis of metal nanoparticles by bacteria has emerged as a promising environmentally-friendly alternative to traditional chemical synthesis. However, this technology remains limited by the lack of knowledge regarding metal reduction and nanoparticle biosynthesis mechanisms within various bacteria, hampering the ability to control nanoparticle properties necessary for various applications. Since biosynthesis is commonly viewed as a stress response mechanism, initiated to detoxify toxic metal ions present in the environment, we investigated the role that RNA regulators such as small RNAs (sRNAs) may play in nanoparticle synthesis and functionality. This study was performed in the context of *Deinococcus radiodurans,* where we created deletion strains of selected sRNAs and find that some of these RNA regulators significantly affect the yield of biosynthetic silver nanoparticles (AgNPs), indicating their importance in nanoparticle biosynthesis. We also demonstrate that modulating the levels of these sRNAs can improve the catalytic and antimicrobial activity of the resulting AgNPs relative to AgNPs synthesized using the wild-type strain. Since functionality is often dictated by the properties of the AgNPs, further materials characterization of the AgNPs reveals drastic changes in the surface chemistry and overall composition of the AgNPs upon deletion of specific sRNAs. Taken together, our results demonstrate how modulating sRNA levels enables the biosynthesis of AgNPs with unique material properties and improved functionality. As such, we introduce sRNAs as a new platform for engineering the biosynthesis of metal nanoparticles using bacteria.

Automated Design of Genetically-Encoded Protein-Sensing Riboswitches in a Cell-Free Expression System

Vezeau, G. E. - Presenter, Pennsylvania State University

Gadila, L. R., Penn State University

Salis, H. M., Pennsylvania State University

A diverse array of genetically-encoded sensor devices have been implemented in cell-free systems, enabling the detection of toxins like fluoride and pathogens like Zika virus in a portable and cost-effective platform. However, protein detection has lagged behind. Here, we use RNA aptamer sensing elements that detect specific proteins to design riboswitches, using a previously-developed model of riboswitch translational regulation, the Riboswitch Calculator. We designed 35 activating and repressing riboswitches, detecting the phage MS2 coat protein, human monomeric C-reactive protein, and human IL32Î³, and used a TXTL cell-free expression system as a test platform to determine riboswitch function. The best ON riboswitch activated the expression of a fluorescent reporter up to 15.9-fold, while the best OFF riboswitch repressed reporter production down to 13.6-fold. We discovered that, while the MS2 and IL32Î³ riboswitches functioned largely due to inducing structural changes in the ribosome binding site, the function of the CRP riboswitches was also affected by direct CRP-mediated repression. We further demonstrate the importance of accurate structural knowledge of aptamers for the precise design of highly-functional riboswitches. Overall, we expand the genetically-encoded sensor toolbox, and demonstrate the validity of automated computational design of protein-detecting RNA structural switches.

Automated Metabolic Rewiring Design for Optimized Succinate Production in Saccharomyces Cerevisiae

Jansen, G. - Presenter, University of Cambridge

Nicosia, G.

Costanza, J., Italian Institute of Technology

Amaradio, M. N., University of Catania

Pappalardo, X., University of Catania

Succinic acid, or *succinate*, is the precursor of a wide range of consumer products and biotechnological applications. Hence, a sustainable microbial bioproduction of succinate by engineered *Saccharomyces cerevisiae* strain is one of the highest markets demands in the large-scale fermentation. In the perspective of novel design for competitive engineered yeast strains, in the present study we propose the application of a powerful pipeline based on non-linear optimization for *in silico* design of metabolic networks through specific gene deletions. The *ad-hoc* bio-inspired algorithm that we have developed uses a genome-scale model reconstruction of the metabolic network in S. cerevisiae and the distribution of metabolic fluxes predicted by *flux balance analysis*. Succinate production was obtained on a wide range of different media including anaerobic medium and with different combinations of carbon sources. Multiple *in silico* simulations of easy genetic manipulation of strains allow the selection of the most competitive ones within the approximated Pareto Fronts. The valuable gene deletions were iteratively and competitively selected through the constrained multi-objective optimization algorithm, which were run for a low number of objective function evaluations, followed by an ad-hoc post-processing policy keeping the significant changes for succinate production. Other constraints were also considered for the refinement of the pipeline, including the variability of the predictions, the allowable reduction on key fluxes within the metabolic network, and the role of essential genes. The resulting strains are scattered in the phenotypic space, with clusters of points showing similar *growth and succinate production*; most competitive strains can then be selected for further studies depending on other characteristics of interest, such as particular by-products or productivity. The potential of computational methodologies in design automation for metabolic engineering offers a powerful approach for improving the effective production of succinate and establishing a more efficient and much less time-consuming in vitro mutagenesis.

Automated Sample-to-Sequencer Library Preparation Using Programmed Centrifugal Force, Pneumatic Force, and Temperature Control

Guo, J. - Presenter, National Research Council Canada

Brassard, D. - Presenter, National Research Council Canada

Veres, T., National Research Council Canada

Mounier, M., National Research Council Canada

Janta-Polczynski, M., National Research Council Canada

Miville-Godin, C., National Research Council Canada

Pilar, A., Health Canada

Corneau, N., Health Canada

Charlebois, D., Canadian Space Agency

Ferreira, J., National Research Council Canada

Precise control of gene expression is critical for biological research and biotechnology. However, transient plasmid transfections in mammalian cells produce a wide distribution of plasmid copy numbers per cell, and consequently, high expression heterogeneity. High expression heterogeneity can make genetic circuits less predictable and unreliable. Here, we report plasmid-based synthetic circuits â€" Equalizers â€" that buffer copy number variation at the single-cell level. Equalizers couple a transcriptional negative feedback loop with post-transcriptional incoherent feedforward control. We demonstrate that Equalizers outperform other gene dosage compensation topologies and produce as low cell-to-cell variation as chromosomally integrated genes. We also show that episome-encoded Equalizers enable a rapid generation of extrachromosomal cell lines with stable and uniform expression for over two months. Overall, Equalizers are simple and versatile devices for homogeneous gene expression and can facilitate the engineering of synthetic circuits that function reliably in individual cells.

Autonomous Cell-Free Synthesis and Assembly of a Ribosomal Subunit

Daube, S. S. - Presenter, Weizmann Institute of Science

Levy, M., Weizmann Institute of Science

Falkovich, R., Massachusetts Institute of Technology

Bar-Ziv, R., Weizmann Institute of Science

Ribosome biogenesis is an efficient and complex assembly process that has not been reconstructed outside a living cell so far, yet is the most critical step for establishing a self-replicating artificial cell. We recreated the biogenesis of *Escherichia coliâ€™s* small ribosomal subunit by synthesizing and capturing all its ribosomal proteins and RNA on a chip[1]. Surface confinement provided favorable conditions for autonomous step-wise assembly of new subunits, spatially segregated from original intact ribosomes. Our real-time fluorescence measurements revealed hierarchal assembly, cooperative interactions, unstable intermediates, and specific binding to large ribosomal subunits. Using only synthetic genes, our methodology is a crucial step towards creation of a self-replicating artificial cell and a general strategy for the mechanistic investigation of diverse multi-component macromolecular machines.

(1) Levy, M.; Falkovich, R.; Daube, S. S.; Bar-Ziv, R. H. Autonomous Synthesis and Assembly of a Ribosomal Subunit on a Chip. *Sci. Adv.* **2020**, *6* (16), eaaz6020. https://doi.org/10.1126/sciadv.aaz6020.

Beyond Fluorescein: Use of Fluorescent Protein Calibrants for Direct and Absolute Quantification of Protein Production in Synthetic Biology

Csibra, E. - Presenter, Imperial College London

Stan, G. B., Imperial College London

While inter-lab calibration standards are approaching mainstream usage in synthetic biology, such calibrations are not in fact sufficient for absolute protein quantification required for modelling synthetic circuits. Fluorescein-based calibration of plate reader and flow cytometry instruments allows the measurement of green fluorescent protein (GFP) in synthetic cells to graduate from arbitrary units to calibrated units, but retains important caveats. Fluorescein is only a good calibrant for green FPs, leaving other FPs uncalibrated, and only provides conversions to units of brightness, not to molecule numbers.

Ideal assay calibrants in molecular biology consist of the same molecule as the one to be measured â€" in this case, a purified preparation of the appropriate fluorescent protein. Here we show that by using purified FP calibrants, all protein species in a synthetic circuit can be quantified in absolute terms using no advanced instrumentation. We develop a SEVA (Standardised European Vector Architecture)-based expression vector that allows the high-level production of soluble protein and describe a straightforward and reliable protocol for the purification of micrograms of FP, followed by a calibration that relates fluorescence activity to protein mass.

We validate this protocol by calculating conversion factors for a panel of commonly-used FPs including superfolder GFP, mCherry, mScarlet-I and mTagBFP2 and showing that the calculated values match well with both FP brightness values as well as protein abundance data from other studies. We apply this tool for the comparison of the detection sensitivity of multiple laboratory instruments, and use these calibrations to debug failing synthetic circuit components. We also demonstrate that the suspected bias of the presence of mCherry on OD600 measurements is real, but in practice requires extreme overexpression (~100,000 proteins per cell) to have a meaningful impact on cell density estimates.

Biocatalytic Synthesis of Human Milk Oligosaccharides Lacto-N-Biose Via Modular Enzymatic Cascade with ATP Regeneration

Du, Z. - Presenter, Shandong University

Liu, Z., Shandong University

Tan, Y., Shandong University

Fang, X., Shandong University

Human milk oligosaccharides (HMOs), the third most abundant solid component of human milk, are reported to be beneficial to infant health (Eriksen et al., 2018; Han et al., 2012). The biosynthesis of Lacto-*N*-biose (LNB), the building block for HMOs, suffers from excessive cofactors addition and intermediate inhibition (Nishimoto and Kitaoka, 2007; Nishimoto 2020). Here, we developed an *in vitro* multienzyme cascade composed of LNB module, ATP regeneration, and pyruvate oxidase-driven phosphate recycling to produce LNB. The integration between ATP regeneration and Pi alleviation increased the LNB conversion ratio and resulted in a Î”G'Â° decrease of 540 KJ/mol. Under the optimal conditions, the LNB conversion ratio was improved from 0.34 to 0.83 mol/mol GlcNAc and the ATP addition decreased to 50%. Finally, about 0.96 mol/mol GlcNAc and 71.6 mg LNB g^{-1} GlcNAc h^{-1} of LNB yield were achieved in 100 mL reaction system. The synergistic strategy not only paves the way for producing LNB but also facilitates other chemicals with multienzyme cascades.

Reference:

Du, Z. Q.; Liu, Z. Y.; Tan, Y. S.; Niu, K. L.; Guo, W.; Jia, Y.Y.; Fang, X. (2021). Lacto-*N*-biose synthesis via a modular enzymatic cascade with ATP regeneration. ***iScience***, 24(3), 102236.

Eriksen, K. G.; Christensen, S. H.; Lind, M. V.; and Michaelsen, K. F. (2018). Human milk composition and infant growth. ***Curr. Opin. Clin. Nutr***. 21 (3), 200-206.

Han, N. S.; Kim, T. J.; Park, Y. C.; Kim, J.; and Seo, J. H. (2012). Biotechnological production of human milk oligosaccharides. ***Biotechnol. Adv***. 30 (6), 1268-1278.

Nishimoto, M. (2020). Large scale production of lacto-*N*-biose I, a building block of type I human milk oligosaccharides, using sugar phosphorylases. ***Biosci. Biotech. Bioch***. 84, 17-24.

Nishimoto, M. and Kitaoka, M. (2007). Practical preparation of lacto-*N*-biose I, a candidate for the Bifidus factor in human milk. ***Biosci. Biotech. Bioch***. 71 (8), 2101-2104.

Bio-Production of Poly-DOPA As a Functional Material and Electron Acceptor for Extracellular Electron Transport and Bioelectrochemical Systems

Walker, D. - Presenter, Univeristy of Texas, Austin

Biopolymers have attracted considerable interest as material coatings that could interface biologics and abiological materials for drug delivery, biosensing, and bioelectronic applications. One such bio-polymer, Poly-DOPA, has recently gained traction in the fields of energy storage and bioelectrochemical systems. However, despite its wide use, little work has been done on engineering L-DOPA production so that cells might create their own Redox-active coatings. We have now developed genetically modified strains of *Escherichia coli* and *Shewanella oneidensis* that produce extremely high concentrations of the monomer L-DOPA. When grown under aerobic condition, the L-DOPA undergoes self-oxidation, polymerizing into PolyDOPA. Liquid media turns black, resulting in a broad UV-Vis absorbance spectra from the supernatant, and on solid media the PolyDOPA permeates through the hydrogel. L-DOPA production and polymerization can be controlled temporally and spatially to create gradients of PolyDOPA, resulting in different compositions and colors within hydrogels. Upon electrodeposition and polymerization of L-DOPA to PolyDOPA on a graphite electrode, we found an increased ability of the bacterial cells to bind to the electrode, which in turn resulted in more rapid establishment of a bioelectrochemical cell, and an overall higher current output. Furthermore, L-DOPA is capable of acting as a soluble, and PolyDOPA, as an insoluble electron acceptor. Taken together, these results suggest a path to a green, self-producing, readily controlled, and (most importantly) scalable material coating that could create more efficient energy production systems.

Bioprospecting in Algal Genomes with Conserved Gene Proximity

Blaby-Haas, C. - Presenter, Brookhaven National Laboratory

McCorkle, S., Brookhaven National Laboratory

Genomes hold the key to unlocking breakthroughs in bioenergy, carbon sequestration, and climate change mitigation. Leveraging insight from natural processes, researchers can design nature-inspired solutions to these challenges. With the successful implementation of an integrated systems biology approach to model, design and engineer high levels of biofuel precursors and value-added products, microalgae have the potential to become major sources of sustainable bioenergy and bioproducts. Toward this goal, over 100 algal whole-genome sequences are either presently available or are soon to be published. Remarkably, over half of the proteins encoded by algal genomes are of unknown function, highlighting both the volume of unique functional capabilities yet to be discovered and a fundamental knowledge gap that limits available design space and impedes successful biosystem design. To address the bottleneck in gene function understanding, we are employing a comparative genomic technique to identify conserved proximal genes. The resulting functional neighborhoods are providing insight into gene function and revealing design principles for the construction of synthetic pathways in algae.

Cas9-Mediated Random Access in DNA Data Storage

Cardozo, N. - Presenter, University of Washington

Zhang, K. - Presenter, University of Washington

Nivala, J., University of Washington

Ceze, L., University of Washington

Strauss, K., University of Washington

Wilde, D., University of Washington

Anderson, C., University of Washington

Synthetic DNA is an excellent medium for long-term digital data storage as it can encode data with high density and durability. One requirement of practical DNA data storage systems is the ability to selectively access data files (a process known as "random access") to lessen decoding latency and cost. We and others have previously explored PCR-addressable access within DNA data pools. However, this process is not ideal due to the additional sample prep, long cycle times, and occurrence of artifacts involved with PCR amplification. Here, we demonstrate a new random access approach in which files can be selected in multiplex using a CRISPR-Cas9 target address and then decoded using a nanopore sequencer. We designed and validated this random access architecture within a pool of 1.6 million DNA sequences representing 25 individual data files. This strategy has several advantages over previous work, including 1) decreased time-to-decoding, 2) increased one-pot multiplexability, and 3) increased energy efficiency due to replacing thermocycling steps with an isothermal reaction.

Cell-Free Detection of Heavy Metals for Rapid, in-Field Water Quality Testing

Bernhards, C. - Presenter, CCDC Chemical Biological Center (Excet, Inc.)

Beabout, K., UES, Inc.

Thakur, M., George Mason University

Turner, K., NRL

Chávez, J., Air Force Research Laboratory

Walper, S., US Naval Research Laboratory

Lux, M., CCDC Chemical Biological Center

The ability to detect contaminants, such as heavy metals, in water is critical to health and safety in both field-forward and resource-poor environments. Nature and bioengineers have devised biosensors that respond to a variety of analytes; however, deployment of these sensors remains a challenge. For cell-based sensors, these challenges include the nutrient requirements of living cells and concerns surrounding the potential release of genetically modified organisms in the environment. Cell-free protein synthesis (CFPS) is a powerful tool that is being harnessed for the development of biosensors due to the ability of CFPS systems to detect an array of analytes, including those that are toxic to living systems, and their potential to address other limitations of whole-cell biosensors. CFPS systems can be freeze-dried onto paper substrates, and following rehydration, a visualizable reporter protein is produced in response to the presence of the target analyte. These paper-based sensors are cheap, disposable, simple to use, and multiplexible. This work shows a tri-service effort to develop cell-free biosensors for arsenic, cadmium, and mercury. Sensor optimization using an acoustic liquid handler and specificity data are also presented. These cell-free sensors are able to detect heavy metal concentrations at or below US Military Exposure Guidelines in under 30 minutes, and we demonstrate the simultaneous detection of multiple metal species in one assay. In addition to heavy metals, the CFPS approach described here can be expanded in the future to detect a wider range of contaminants, including toxic compounds and pathogens.

Characterizing microRNA-Mediated Modulation of Gene Expression Noise and Its Effect on Synthetic Gene Circuits

Wei, L. - Presenter, Tsinghua university

Li, S., Tsinghua University

Wang, X., Building of biotechnology

Gene expression noise plays an important role in many biological processes, such as cell differentiation and reprogramming. It can also dramatically influence the behavior of synthetic gene circuits. MicroRNAs (miRNAs) have been shown to reduce the noise of lowly expressed genes and increase the noise of highly expressed genes, but less is known about how miRNAs with different properties may regulate gene expression noise differently. Here, by quantifying gene expression noise using mathematical modeling and experimental measurements, we showed that competing RNAs and the composition of miRNA response elements (MREs) play important roles in modulating gene expression noise. We found that genes targeted by miRNAs with weak competing RNAs show lower noise than those targeted by miRNAs with strong competing RNAs. In addition, in comparison with a single MRE, repetitive MREs targeted by the same miRNA suppress the noise of lowly expressed genes but increase the noise of highly expressed genes. Additionally, MREs composed of different miRNA targets could cause similar repression levels but lower noise compared with repetitive MREs. We further observed the influence of miRNA-mediated noise modulation in synthetic gene circuits which could be applied to classify cell types using miRNAs as sensors. We found that miRNA sensors that introduce higher noise could lead to better classification performance. Our results provide a systematic and quantitative understanding of the function of miRNAs in controlling gene expression noise and how we can utilize miRNAs to modulate the behavior of synthetic gene circuits.

Construction of a Modular Framework for Orthogonal and Predictable Gene Expression in Prokaryotes

Kar, S. - Presenter, University of Texas at Austin

Ellington, A., University of Texas at Austin

Gollihar, J., US Army Research Laboratory- South

Wang, C., University of Texas at Austin

Mcleland, C., University of Texas at Austin

Synthetic gene circuits can produce complex biomolecules and implement complex computations, but their performance is often limited to model organisms. By incorporating orthogonal transcriptional control elements in the design and construction of synthetic circuits it should prove possible to increase robustness, predictability, and portability across diverse organisms, while mitigating potential crosstalk and other failure modes. In prokaryotes, orthogonal control has often been achieved with the use of orthogonal polymerases (RNAPs), in particular T7 RNAP. Unfortunately, T7 RNAP-mediated expression frequently leads to a drain of cellular resources and associated toxicity, and there is a need for more precise control over such systems. Here we present a universal plasmid system (UPS) for the modular and facile construction of synthetic genetic systems based on orthogonal transcriptional control elements, and simple â€˜macrosâ€™ that lead to different states of T7 RNA polymerase-based expression. Initially, we have focused on T7 RNAP homeostasis circuits, which produce defined gene expression under a variety of conditions, since both the production and regulation of T7 RNAP are under its own control and insulated against host RNAP regulation. By systematically assembling versions of the homeostasis circuit, we demonstrate that in a variety of *Escherichia coli* strains â€“ DH10B, MG1655 and the probiotic strain Nissle 1917 â€“ gene expression outputs can be predicted by a simple quantitative model and readily controlled via mutant T7 RNAP promoters. Furthermore, by encoding all circuit elements on a single broad host range vector, the UPS homeostasis system can be ported across a number of industrially and therapeutically relevant species â€“ *Pseudomonas putida, Salmonella typhimurium, Serratia marcescens*, and *Bacillus subtilis* â€“ and can be used to constitutively express genes at preset levels based solely on T7 RNAP-mediated control. Further instantiations of additional T7 RNAP-based â€˜macrosâ€™ should allow the expression of spike, oscillatory and other waveforms of expression.

Construction of Synthetic Human Antibody Library Based on Repertoire Sequencing

Katoh, H. - Presenter, Tokyo Medical and Dental University

Ishikawa, S., Tokyo Medical and Dental University

Komura, D., Tokyo Medical and Dental University

Human monoclonal antibody is a good modality of molecular targeting therapy for various diseases including inflammatory dysregulations and malignant tumors. Recent advances in immune-repertoire sequencing have enabled us to identify disease-related human antibodies as well as candidate therapeutic antibodies for malignancies from disease specimens (#1-3). By analyzing immunoglobulin repertoires of tumor infiltrating B-lymphocytes in human cancers and synthesizing antibodies based on the sequencing information, we discovered that approximately one-third of dominant B-lymphocyte clones in tumors expressed antibodies that reacted to sulfated glycosaminoglycans (#1), and another one-third expressed those corresponding to ribonucleoproteins in the focal adhesion complex (#2). Moreover, the anti-sulfated glycosaminoglycan antibodies found in human disease specimens have been revealed to be candidate therapeutic antibodies. Thus, the sequencing information of the repertoire sequencing by itself can be precious database to develop therapeutic human antibodies.

However, although next generation repertoire sequencing acquires hundreds of thousands of immunoglobulin repertoires from a single specimen, methods for the construction of disease-specific antibody library, its amino acid modification strategy, and functional screening has not been established to date. Here, based on repertoire information, we show a high-throughput method for the selection and synthesis of cDNA library of disease-related human antibodies and a rational designing strategy of amino acid modifications to obtain higher functional human monoclonal antibodies.

#1. Katoh H, et. al., Immunogenetic profiling for gastric cancers identifies sulfated glycosaminoglycans as major and functional B cell antigens in human malignancies. Cell Rep 2017.

#2. Konishi H, et. al., Capturing the differences between humoral immunity in the normal and tumor environments from repertoire-seq of B-cell receptors using supervised machine learning. BMC Bioinformatics 2019.

#3. Atsumi S, et. al., Focal adhesion ribonucleoprotein complex proteins are major humoral cancer antigens and targets in autoimmune diseases. Commun Biol 2020.

Contact-Free Detection of Sars-Cov-2 Via a Cell-Free Circuit to Resonant Sensor Interface

Dopp, J. - Presenter, Iowa State University

Carr, A., Iowa State University

Wu, K., Boston University

Sadat Mousavi, P., University of Toronto

Jo, Y. R., Iowa State University

Pardee, K., University of Toronto

Green, A. A., Arizona State University

Reuel, N., Iowa State University

The ongoing COVID-19 pandemic has highlighted the need for timely, cost-effective, and accurate diagnostics. It has also revealed how shortages of plastic consumables, buffers, and PPE can limit traditional assays. Sample collection and centralized testing with RT-qPCR has dramatically improved the time to result, but it remains too slow to mitigate the spread of infection. To help address these issues, we have developed a frugal SARS-CoV-2 diagnostic enabled by cell-free synthetic biology circuitry coupled to a paper-based resonant sensor that can be safely and discretely scanned in a closed envelope and has the potential for at-home use and distributed contact-free testing. Lysed samples undergo isothermal reverse transcription recombinase polymerase amplification (RT-RPA) in a sample well which is placed onto a lyophilized, paper-based cell-free expression system. The cell-free system contains plasmids that harbor toehold switches selective to specific targets on the amplified RPA products. Complementary viral RNA amplicons complex with the toehold switches to activate translation of subtilisin, a strong protease reporter. The protease then degrades gelatin switches coated on the paper, which overlay a screen-printed paper-based resonant (LC) sensor. The sensor has a resonant frequency which shifts due to changes in permittivity caused by gelatin degradation. This clear frequency change can be measured rapidly through a closed envelope using an embedded portable vector network analyzer. We envision the technology can be deployed with autonomous scanners positioned at local drive-up lanes, hospitals, or workplaces and would be capable of signaling within 10 seconds if the closed sample envelope contains a positive or negative sample. This workflow of individualized administration followed by rapid, local scanning would eliminate the current problems of test site back-log, lack of PPE, and reagent scarcity and would be well-suited for deployment against a variety of existing and emerging pathogens for outbreak tracking and pandemic preparedness.

Crage Enables Rapid Activation of Biosynthetic Gene Clusters in Undomesticated Bacteria

Yoshikuni, Y. - Presenter, Lawrence Berkeley National Laboratory

It is generally believed that exchange of secondary metabolite biosynthetic gene clusters (BGCs) among closely related bacteria is an important driver of BGC evolution and diversification. Applying this idea may help researchers efficiently connect many BGCs to their products and characterize the productsâ€™ roles in various environments. However, existing genetic tools support only a small fraction of these efforts. Here, we present the development of chassis-independent recombinase-assisted genome engineering (CRAGE), which enables single-step integration of large, complex BGC constructs directly into the chromosomes of diverse bacteria with high accuracy and efficiency. To demonstrate the efficacy of CRAGE, we expressed three known and six previously identified but experimentally elusive non-ribosomal peptide synthetase (NRPS) and NRPS-polyketide synthase (PKS) hybrid BGCs from *Photorhabdus luminescens* in 25 diverse Î³-Proteobacteria species. Successful activation of six BGCs identified 22 products for which diversity and yield were greater when the BGCs were expressed in strains closely related to the native strain than when they were expressed in either native or more distantly related strains. Activation of these BGCs demonstrates the feasibility of exploiting their underlying catalytic activity and plasticity, and provides evidence that systematic approaches based on CRAGE will be useful for discovering and identifying previously uncharacterized metabolites.

Creation of CO2 Emission Suppressed E. coli By Combination of CO2 Transporter/Fixation-Enzyme Introduction and Directed Evolution

Nishida, H. - Presenter, Hitachi, Ltd.

Nakazawa, S., Hitachi, Ltd.

Kato, M., Hitachi, Ltd.

Ohta, M., Hitachi, Ltd.

Ito, K., Hitachi, Ltd.

Tanabe, M., Hitachi, Ltd.

Effective CO_2 fixation from the air is urgent issue for preventing global climate change. Especially the CO_2 transformation into high value-added chemicals by synthetic biology is considered to be one of the promising solutions with economic viability. In the synthetic biology, *Escherichia coli* is widely utilized as a platform cell because of its established genetic engineering methods and possible high-density culture. We constructed the genetically engineered *E.coli* cell transformed with CO_2 transporter (bicA) and phosphoenolpyruvate carboxykinase (pck) for the subsequent CO_2 fixation process. On the other hand, we applied the wild type to the directed evolution process, in which UV-irradiated *E. coli* cells were cultured under high CO_2 environment and survived clones were selected. We evaluated the characteristics of those cells by our created measurement system for the CO_2 concentration in sealed container. As a result, the engineered (bicA/pck) and evolved (high CO_2) cells exhibited lower CO_2 emissions than that from the wild type. When the evolved cells were transformed with bicA and pck, those cells showed more suppressed CO_2 emission than the engineered and the original evolved cells. Both engineered and evolved cells showed lower CO_2 emission than that from the wild type, suggesting that the both cells incorporated CO_2 to some extent, and assimilated CO_2 to the other metabolite in the cells. In order to elucidate the genetic mechanism behind, genetic analysis of evolved cells are now under investigation. We would like to report our preliminary study on our directed evolution method and its effectiveness in creating novel characteristics of bacterial cells.

CRISPR/Cas9 Allele-Specific Targeting of Autosomal Dominant Retinitis Pigmentosa Disease Variants

Geiger, A. - Presenter, The University of Adelaide

Thomas, P., The University of Adelaide

Godahewa, G., CSIRO

Adikusuma, F., The University of Adelaide

Wilson, L., CSIRO

Casson, R., The University of Adelaide

Autosomal dominant Retinitis Pigmentosa (adRP) is an important cause of progressive, irreversible blindness. Clinically and genetically heterogeneous, adRP is known to be caused by a myriad of heterozygous variants in at least 29 genes. Where adRP is caused by point mutations in haplosufficient genes with dominant-negative and/or toxic gain-of-function disease mechanisms, selective ablation of the mutant allele using CRISPR/Cas9 technology represents a tantalising prospect for treatment. The Pro23His mutation in Rhodopsin (*RHO*) and the Gly56Arg mutation in Nuclear Receptor Subfamily 2 Group E Member 3 (*NR2E3)* are particularly attractive targets, representing the two most common causes of non-syndromic adRP, respectively. However, guide RNA design is heavily restricted due to minimal availability of â€" or in the case of *RHO* Pro23His, the absence of â€" canonical SpCas9 5â€™-NGG-3â€™ PAM sequences in these regions. In this study, we have explored the application of alternative naturally occurring or engineered CRISPR/Cas9 platforms to target both the *RHO* Pro23His and *NR2E3* Gly56Arg mutations. To assess disease-allele targeting activity we generated clonal cell models of both mutations using a novel PRIME editing construct. Analysis of gene editing outcomes identified several candidate gRNAs with selective targeting of the mutant allele, with a prevalence of frame-shifting indels or larger deletions observed. These data provide evidence that our approach has the potential to mediate specific gene editing events that will ablate the mutant allele whilst sparing the wild-type allele, positing these strategies as therapeutic candidates for adRP.

Deep Mutational Scanning (DMS) Guided Therapeutic Enzyme Engineering

Nair, N. - Presenter

Phenylalanine ammonia-lyases (PALs) non-oxidatively deaminate L-phenylalanine to *trans*-cinnamic acid (tCA) and are widely found associated with secondary metabolism in plants, bacteria, and fungi. Biocatalytic applications for natural product and fine chemical synthesis has driven the discovery, expression, characterization, and engineering of PALs. More recently, development of PALs for phenylketonuria (PKU) management and cancer therapy has further increased interest in engineering this class of enzymes. While there is a general understanding of how residues in the substrate-binding pocket contribute to specificity and turnover, led by rational mutagenesis studies, there is poor understanding of how distal residues affect function. Deep mutational scanning (DMS) can provide a comprehensive map of sequenceÃ¢â‚¬â€œfunction relationships to explore the protein fitness landscapes, uncover functionally relevant sites, and improve molecular energy functions. Yet, DMS has not been widely coupled with directed evolution to accelerate engineering campaigns (although there are few exceptions).

Previously, we developed a growth-coupled enrichment for rapid screening of high-activity variants of AvPAL* (used to formulate the PKU drug PegvaliaseÂ®) in *E. coli*. Here, we first obtained the detailed sequence-function landscape of PAL, to date, using DMS, identifying >60 mutational hotspots. Next, we picked seven sites for comprehensive single and multi-site saturation mutagenesis and we identified multi-site mutations with improved activtiy. We then explored the epistatic effect of these mutations, uncovering positive, neutral, and negative interactions among distal and proximal sites. Finally, to understand the mechanistic role of key mutations in hyperactive variants, we performed modelling studies (quantum mechanics, molecular dynamics, including metadynamics) and concluded that there are multiple pathways to enhance PAL catalytic activity. In summary, this study demonstrates that DMS can guide and expedite enzyme directed evolution and campaigns as well as mechanistic studies.

Designing Synbio Products with the End in Mind – Techno-Economic Analyses (TEAs) As a Tool for Strategic Development

Crumbley, A. M. - Presenter, The University of Alabama

Recent efforts to generate biomaterials and chemical precursors through federal programs, such as DARPA 1000 Molecules, have created vast libraries of microbes capable of producing both essential and novel SynBio products. Army SynBio products have the potential to revolutionize everything from essential supply chains to device performance, both supporting the warfighter in the field and creating opportunities for public sector spin-off technologies. Considering the impact of physical, technical, and economic scale-up factors during strain engineering, including physical limits imposed by process capabilities at larger scales, byproduct impact on downstream separation strategies, and feedstock market factors, can substantially impact successful transition of SynBio products from the lab bench to the industrial facility. One tool to strategically assess a SynBio project in development is the techno-economic analysis (TEA). To illustrate the insight that early-stage logistics consideration can offer for SynBio product design, we performed a retrospective TEA for a recent in-house scale-up operation. *E. coli* filamentous phage-based BioCNF is a carbon filtration material with properties tunable through bioengineering of the phage template. BioCNF is useful for oxygen storage, gas mask filtration, and fiber-based technologies, and was recently produced at the 10 lb (4.5 kg) scale using DEVCOM CBCâ€™s pilot scale BioManufacturing facility. TEA insights suggest transitioning the strain to a lower-cost media would improve overall economics.

Developing a Novel Microbial Chassis for Upcycling Waste Polyethylene Terephthalate

Moon, T. S. - Presenter, Washington University in St. Louis

Diao, J., Washington University in St. Louis

Hu, Y., Washington University in St. Louis

Polyethylene terephthalate (PET) represents 8% of global solid waste. PET chemical recycling has been an option to solve this global problem, but it suffers from its relatively high process cost and the extremely low price of virgin PET. One solution to address this issue is to upcycle waste PET rather than recycle it to generate the same PET typically with low quality. PET upcycling can be achieved by depolymerizing PET into terephthalic acid (TPA) and ethylene glycol (EG) and biologically converting these monomers into value-added products. However, there are only a handful of reports demonstrating microbial strains capable of growing on both TPA and EG generated from PET as sole carbon sources. To overcome this critical limitation, we have performed strain screening to discover a *Rhodococcus* strain (named RPET) that can grow well on the alkaline hydrolysis products of PET as the sole carbon source without any purification step. Notably, this strain was able to tolerate and grow on a mixture of TPA and EG at extremely high concentrations (up to 0.3M each, total 0.6M) and high osmolarity resulting from alkaline hydrolysis and pH neutralization. The resultant pH neutralized media supported RPETâ€™s growth (up to 0.4 g dry cell weight per g PET) without any purification and sterilization step except for their dilution to make up to 0.6M of monomer concentrations. In addition, many synthetic biology tools, developed for a related species *Rhodococcus opacus* (1), were functional in RPET, facilitating its engineering. In this presentation, we will discuss our effort to develop this novel chassis for waste PET valorization with PET conversion into carotenoids and muconate as two demonstration products (2).

(1) DeLorenzo DM, Rottinghaus AG, Henson WR, Moon TS. 2018. *ACS Synthetic Biology* 7:727-38

(2) Moon TS et al. U.S. Provisional Patent Application

Developing the Chlamydomonas Reinhardtii Plastid As a Platform for Directed Evolution of Photosynthetic Proteins

Stainbrook, S. C. - Presenter, Northwestern University

Daki, A., Illinois Math and Science Academy

Ummareddy, R., Illinois Math and Science Academy

Qu, K., Illinois Math and Science Academy

Although plants are essential to human life and industry, their photosynthetic properties are not perfectly efficient. Much effort is being invested to improve photosynthesis, especially crop species. These efforts are often hampered by the plantsâ€™ genomic intractability, large space requirements and long generation time. We propose *Chlamydomonas reinhardtii*, the â€œgreen yeast,â€ as a new platform for high throughput engineering of photosynthetic proteins.

C reinhardtii is a haploid, unicellular alga that has already been developed as a model organism for study of photosynthesis because of its close homology to land plants. Proteins from plants functionally complement the knockout of their algal homologs. The plastome is readily transformed, allowing photosynthetic proteins to be expressed in their native chloroplast environment. Additionally, homologous recombination in the plastome allows protein variants to be readily compared because identical genetic context eliminates the confounding effects traditionally seen with random genomic integration. Because of the algaâ€™s small size and rapid growth, protein variants can be assayed on a scale not feasible in plants. When a plant protein is used to complement an algal knockout, the phenotype will be determined by the plant protein so that the characteristics of that protein can be selected directly. In this way, high throughput directed evolution of plant proteins can be realistically achieved. When improved protein variants are found and characterized in the algal system, they will be validated in land plants.

We present initial progress to create a Golden Gate system for integration and expression of plant genes from the *C reinhardtii* plastome. The protein Rubisco Activase has been chosen for proof of concept because its lack of thermal stability is recognized as a key limitation for heat tolerance in C4 crops such as maize.

Development of a Synthetic Biotic, SYNB8802, for the Treatment of Enteric Hyperoxaluria

Lubkowicz, D. - Presenter, Synlogic

Renaud, L., Synlogic

Reeder, P. J., Synlogic Inc

Ghoshal, S., Synlogic

Horvath, N.

Bergeron, C., Synlogic

Chen, C., Synlogic

James, M., Synlogic Inc.

Cantarella, P., Synlogic

Charbonneau, M., Synlogic Inc

Shmueli, R., Synlogic

Gao, J., Synlogic

Puurrunen, M., Synlogic

Isabella, V., Synlogic Inc.

Perreault, M., Synlogic

Enteric hyperoxaluria (EH) is a human metabolic disease that results from excessive absorption of oxalate from dietary sources. High urinary oxalate levels (UOx) in patients with enteric hyperoxaluria (EH) can lead to recurrent kidney stones, nephrocalcinosis and chronic kidney disease (CKD). EH disproportionately impacts patients who have undergone certain kinds of gastric bypass surgery and can lead to significant kidney complications in those patients. It also occurs in patients with underlying GI disorders including Crohnâ€™s disease or short bowel syndrome. There is a high unmet need for new EH therapies as there are currently no approved pharmacological therapies for treating hyperoxaluria. Synlogic is developing a novel Synthetic Biotic™ medicine for the treatment of EH, SYNB8802. SYNB8802 is an engineered bacterium derived from *Escherichia coli* Nissle 1917 (*EcN*) that has been engineered to metabolize oxalate to formate and CO_2 in the gastrointestinal tract. When SYNB8802 was administered concomitantly with ^{13}C-oxalate to healthy mice, it decreased the urinary ^{13}C-oxalate, which is indicative of its ability to consume oxalate *in vivo*. In healthy non-human primates (NHP) administered ~400 mg of oxalate, SYNB8802 dose-dependently lowered the urinary oxalate and ^{13}C-oxalate by up to 75% as compared to vehicle. In addition, Synlogic has developed a mathematical model that is predicting clinically meaningful reductions in urinary oxalate in humans. SYNB8802 is currently in Phase 1 clinical trials in healthy volunteers and patients with EH secondary to Roux-en-Y gastric bypass surgery. Overall, SYNB8802 represents a promising new approach for the treatment of enteric hyperoxaluria.

Development of a Toggle Switch Based on Post-Transcriptional Regulation That Resists Epigenetic Silencing

Wauford, N. - Presenter, MASSACHUSETTS INSTITUTE OF TECHNOLOGY

DiAndreth, B., MIT

Weiss, R., Massachusetts Institute of Technology

Synthetic biological circuits may enable sophisticated cell therapies that respond to patient biomarkers and inputs from clinicians. However, transcriptionally regulated genetic circuits are prone to epigenetic silencing, in which chromatin modifications often turn off transgene expression after a few weeks. This poses a significant challenge to the regulation of cell therapies that may persist *in vivo* for years. Constitutive transcription has been shown to reduce silencing, but tools to regulate gene expression post-transcriptionally are limited. To address this gap, we present the PERSIST platform for post-transcriptional regulation, in which RNA cleavage events activate or repress downstream gene expression. We take advantage of CRISPR proteins in the Cas6 and Cas13 families that cleave sequence-specific RNA hairpins, demonstrating a set of nine proteins that enact activation and repression, often with fold changes of two orders of magnitude. PERSIST genetic circuits show 28% less silencing than the transcriptionally regulated TetON system after 1 month and no more silencing than a constitutive control. We then used the PERSIST platform to implement a genetic toggle switch. Toggle switches could improve cell therapies by imparting them with memory, such that a transient input, such as a drug input from a clinician, can permanently change cell behavior.

We designed a system in which FDA-approved small molecule drugs could switch a cell between two behavioral states, and implemented this system by combining PERSIST logic processing with engineered small molecule-responsive fusion proteins. Preliminary results suggest that our circuit is able to switch between two stable states in response to clinically relevant small molecule drugs. These results will provide a platform to regulate the expression of transgenes in cell therapies over time, enabling optimized dosing schedules of therapeutic proteins in applications including encapsulated cell therapies, engineered immune cells, and gene therapies.

Development of an In Vivo Method in Nepeta Cataria (catnip) to Characterize the Biosynthetic Pathway behind Nepetalactone Stereoisomer Production

Palmer, L. - Presenter, Max Planck Institute For Chemical Ecology

Yamamoto, K., Max Planck Institute for Chemical Ecology

O'Connor, S., Max Planck Institute For Chemical Ecology

Sonawane, P., Max Planck Institute for Chemical Ecology

Chuang, L., Leibniz University Hannover

Siegmund, M., Max Planck Institute for Chemical Ecology

The Lamiaceae plant family, colloquially known as the mint family, is well known for its chemical diversity and economical importance, especially amongst members of the sub-family the Nepetoideae. Most members of this sub-family are well known for their diverse terpene-based natural products; however, one genus, the *Nepeta*, stands out amongst the Nepetoideae for its unique ability to produce nepetalactone, an iridoid-scaffold compound known for its psychoactive effect on cats and potential use as a bio-based pest control in agriculture due to its influence on various insect species. Chemical profiling on *Nepeta* spp. and within varieties of a single species have revealed the production of different nepetalactone stereoisomers varies widely across plants. Previous work has identified *Nepeta* spp. biosynthetic enzymes that can synthesize different stereoisomers of nepetalactones *in vitro*. The role these biosynthetic genes play *in planta* to synthesize different stereochemical ratios of nepetalactone has not been addressed. I developed a virus-induced gene silencing (VIGS) tool for *N. cataria* to explore the *in vivo* function of the putative biosynthetic genes. By simultaneously targeting a visual marker gene, magnesium chelatase subunit H (CHLH), and each member of the genes involved in the production of the various nepetalactone stereoisomers, I am able to precisely select the tissue under the knockdown phenotype of VIGS and characterize this pathway *in vivo*. Furthermore, using this tool, along with expression analysis tools such as qPCR and RNAseq, I aim to untangle the mechanisms behind isomer regulation and gene expression in nepetalactone production, as well as to understand the effect of this pathway on other physiological processes.

Digital Genome Engineering: Unlocking the Unlimited Potential of Biology

Krishnamurthy, N. - Presenter, Inscripta

Abbate, E., Inscripta

Brooks, A., Inscripta

Krouse, K., Inscripta

Leland, B., Inscripta

Rosains, J., Inscripta

Shepherd, T., Inscripta

CRISPR-based genome engineering has the potential to accelerate discovery. Unfortunately, current approaches suffer from limitations in scalability, diversity of edit types, and accessibility. The Onyx™, a benchtop genome engineering platform overcomes these limitations by providing the capability to generate up to 10,000 precisely edited and trackable strain variants in *E. coli* or up to 6,000 such variants in *S. cerevisiae*, in a single run. This enables novel approaches to engineering of gene, pathway, and genome-wide targets in *S. cerevisiae* and *E. coli*. The platform simplifies the complex editing workflow from design to engineered cell library, by including software, reagents, benchtop instrument and analytics. We present 3 applications including high-throughput target discovery, metabolic engineering and accelerated protein optimization. In an application to identify new gene targets that improve glycerol utilization in budding yeast we generated 7 libraries (>40,000 designs) including truncation of non-essential genes and genome-wide expression modulation with terminator or transcription factor binding site modifications. Using barcodes we identified >100 novel targets with a glycerol utilization phenotype. In a metabolic engineering application, we generated 200,000 designs across 24 genome-edited libraries in *E. coli* and screened the resulting strain variants for improved lysine production. We uncovered more than 15 novel genes outside of the known lysine pathway that contribute to increased lysine production. Rapid recombination of the hits through the Onyx workflow resulted in strains with >10,000 fold increase in lysine. To accelerate protein optimization we generated a saturation mutagenesis library of 723 designs against the active site of a genome-integrated GFP and identified several engineered variants showing distinct intensity and spectral characteristics. The Onyx™ platform will usher in the next era of genomics, enabling the researcher to transition from primarily observational studies to truly engineering biology. This will have far-reaching benefits for biology, bio-industrial science, agriculture, healthcare, and alternative energy.

Directed Evolution of a Generalist Biosensor Accelerates Alkaloid Pathway Engineering

d'Oelsnitz, S. - Presenter, UT Austin / Purchasing Office

Ellington, A., University of Texas at Austin

Javanmardi, K., University of Texas at Austin

Thyer, R., University of Texas at Austin

Alper, H., The University of Texas at Austin

In the past decade microbial engineering for production of complex therapeutic plant metabolites has significantly advanced. However, a key bottleneck in the engineering process is screening to identify variants with improved activity, which is typically performed using low-throughput chromatography-based methods. Genetic biosensors can overcome this limitation and increase throughput by several orders of magnitude, but few biosensors exist in Nature for plant metabolites with therapeutic potential. We address this gap by synergizing the extreme promiscuity of a multidrug resistance regulator with a custom directed evolution scheme to create a series of highly specific biosensors for the plant alkaloids tetrahydropapaverine, papaverine, glaucine, rotundine, and noscapine. High resolution structures of these biosensor variants elucidate key adaptations acquired during evolutionary specialization. We subsequently apply one biosensor to evolve a plant methyltransferase, enabling the first microbial production of tetrahydropapaverine, an immediate precursor to four modern pharmaceuticals. Previously studied as virulence factors, multidrug resistance regulators can be repurposed as high throughput analytical tools to accelerate pathway engineering for therapeutic plant metabolites.

Directed Evolution of Synthetic E. coli Predator Population Using Microfluidic Emulsion Droplet Platform

Ganiga Prabhakar, R. - Presenter, Rice University

Shamoo, Y., Rice University

Bennett, M. R., Rice University

Alnahhas, R., Rice University

Hirning, A., Rice University

Microbes have evolved strategies to adapt to nutrient limitations and environmental stresses like competition within microbial communities. Because of these interactions, bacteria have accumulated the ability to produce various specialized metabolites under specific environmental conditions. Activation of these cryptic pathways can provide us with insights into how microbial communities interact as well as access to potentially novel bioactive metabolites for industrial and biomedical use. Here, we use emulsion droplets as a culturing and screening platform to mimic the *in vivo* predator-prey ecosystem by encapsulating single cells from two competing *Escherichia coli* populations.

As a proof of principle, we engineered a model predator-prey system in *E. coli*, such that each droplet contains a Ã¢â‚¬Ëœdefective-predatorÃ¢â‚¬â„¢ and a Ã¢â‚¬ËœpreyÃ¢â‚¬â„¢. A defective-predator has a single nucleotide polymorphism (SNP) to knockout the production of quorum sensing (QS) molecules. An evolved predator activates the engineered cryptic pathway to produce QS molecules required for killing the prey. Therefore, an evolved predator can grow to a higher density compared to a defective-predator within the droplets. By iterating the growth of evolved predators in the emulsion, we could increase its population density to detectable level.

Currently, we have optimized the platform for efficient droplet production, incubation and droplet sorting. The sorting system can detect even a single GFP-labelled *E. coli* cell (pSB1C3_K608010) encapsulated within a 50Âµm droplet. This enables us to selectively screen for functional Predator from a mixed population even at very low densities, compared to a suspension culture. Next, this competition based directed evolution strategy will be applied to harness the SNP reversion in a mutagenized defective-Predator population to produce the quorum sensing molecules under defined co-culture conditions. If successful, this technology can be applied to non-model organisms such as *Streptomyces* to evolve and screen for variants that produce specialized metabolites with bioactive properties.

Dynamic Metabolic Control Improves NADPH Flux and Xylitol Biosynthesis in Engineered E. coli

Li, S. - Presenter, Duke

Lynch, M., Duke University

Metabolic engineering has broad applications in the biosynthesis of a wide variety of products, including biofuels, pharmaceuticals, and food chemicals. Xylitol is a sugar alcohol with a primary use as a sweetener and is produced at ~125,000 tons annually. Compared with the chemical synthesis method, biosynthetic production of xylitol has the potential to decrease costs and be more environmentally friendly. Xylitol can be synthesized from xylose via an NADPH dependent reductase. In our work, we reported improved NADPH flux and xylitol biosynthesis in engineered *E. coli* using 2-stage dynamic metabolic control, where products are made in a metabolically productive phosphate depleted stationary phase. The implementation of this approach relies on the combined use of controlled proteolysis and gene silencing, using degron tags and CRISPR interference, respectively. Strains with reduced levels of enoyl-ACP reductase and glucose-6-phosphate dehydrogenase, led to altered metabolite pools resulting in the activation of the membrane bound transhydrogenase and an NADPH generation pathway, consisting of pyruvate ferredoxin oxidoreductase coupled with NADPH dependent ferredoxin reductase, leading to increased NADPH fluxes, despite a reduction in NADPH pools. These strains produced titers of 200 g/L of xylitol from xylose at 86% of theoretical yield in instrumented bioreactors. We expect dynamic control over the regulation of the membrane bound transhydrogenase as well as NADPH production through pyruvate ferredoxin oxidoreductase to broadly enable improved NADPH dependent bioconversions or production via NADPH dependent metabolic pathways.

Keywords: Xylitol, NADPH, Metabolic engineering, Biosynthesis, Dynamic metabolic control

Reference:

1. Li et al., *Metabolic Engineering* 2021, 64:26-40.
2. Burg et al., *Current Opinion in Chemical Engineering* 2016, 14:121-136

Engineered Probiotics Enabling Specific Sensing of Neurochemicals

Moon, T. S. - Presenter, Washington University in St. Louis

Rottinghaus, A., Washington University In St. Louis

Xi, C., Washington Univeristy In St Louis

Amrofell, M., Washington University in St. Louis

Yi, H., Washington University in St. Louis

Microbes have evolved diverse sensor systems that detect metabolites of great medical importance. These sensors have the potential to be utilized in probiotic microbes to provide diagnostic information and deliver therapeutics with temporal and geographical precision. However, microbial sensors found in nature often have promiscuity to several structurally similar aromatic amino acids, common neurotransmitters, or neuromodulators, limiting their practical applications. Due to differences in associated diseases and functions, specific sensing of these chemicals is critical for probiotic sensor applications. Despite advances in protein engineering for specific ligand-protein interaction, however, engineering ligand-specific sense-and-respond systems remains challenging, especially when the target ligands are structurally similar and ligand-protein binding should control downstream functions such as gene expression. This is mainly due to the challenge in coupling subtle protein conformational changes caused by binding of similar ligands with differential DNA interactions.

We first characterized three promiscuous sensors that recognize aromatic metabolites associated with various metabolic and neurological disorders. Protein engineering often requires extensive structural knowledge of the proteins or massive library sizes. In contrast, to improve ligand selectivity, we rationally engineered the responsible regulators by identifying and individually mutagenizing specific amino acids. We show that our simple and generalizable method of protein engineering is effective and time-efficient, requires small library sizes with only a basic understanding of the protein structure, and enables changes in ligand-protein binding specificity while maintaining protein-DNA interaction and thus downstream gene expression control. Importantly, these sensors are the first ligand-specific sensors for phenylalanine, tyrosine, and phenylethylamine. In addition, our computational and experimental analyses provide novel insights into the uncharacterized regulator structure for the first time, suggesting residues that can be mutagenized to generate novel ligand-specific sensors for additional aromatic neurotransmitters and neuromodulators. The novel ligand-selective sensors generated in this work will have diverse biomedical applications.

Engineering a Global Regulatory System Potentiates Rapid Growth of Saccharomyces Cerevisiae on Numerous Non-Native Substrates

Sullivan, S. F. - Presenter, Tufts University

Trivedi, V., Tufts University

Endalur Gopinarayan, V., LifeMine Therapeutics

There have been decades of effort to engineer the industrial cellular factory *Saccharomyces cerevisiae* to utilize cheap and renewable sources of carbon and energy such as lignocellulosic biomass, glycerol, and various C1 compounds. However, the implementation of heterologous catabolic pathways to enable the assimilation of these non-native substrates has generally relied on constitutive gene expression without consideration for global regulatory systems that may enhance nutrient assimilation and cell growth. Our lab recently demonstrated that activating the galactose (GAL) regulon (a REG approach), the regulatory structure responsible for coordinating growth on the native substrate galactose, during growth on the non-native sugar xylose results in higher growth rates and more efficient substrate utilization compared with traditional constitutive expression (a CONS approach). The present study seeks to investigate whether the large-scale gene expression changes observed upon activation of the GAL regulon are broadly beneficial to growth on non-native substrates. We start by constructing and characterizing a self-activating variant of the sensor-protein (*GAL3-SA*) and find that expression of this mutant is sufficient to fully activate the GAL regulon without an inducing substrate. We find that despite moving from substrate-specific activation (in the native GAL regulon) to non-specific activation, cells carrying *GAL3-SA* do not seem to incur serious fitness costs or sacrifice the previously observed benefits during growth. Finally, we compare CONS and REG approaches to growth on the renewable and structurally distinct substrates arabinose, cellobiose, and glycerol and observe that while a REG approach is superior in all cases, the magnitude of the benefit is both substrate- and media-specific. Ongoing work seeks to identify gene expression patterns that underlie the observed phenotypic benefits across substrates. These results validate GAL regulon activation as a general approach to engineering more efficient catabolism of non-native substrates and provide a tool (*GAL3-SA*) for applying it to other substrates.

Engineering a Programmable Split-Ribozyme Platform for Detecting RNA in Living Cells

Gambill, L. - Presenter, Rice University

Chappell, J., Rice University

Intracellular RNA detection technologies represent a powerful, yet relatively underexplored, area of synthetic biology. With the detection of viral infection, disease states, and antibiotic resistance critical to our ability to combat some of the greatest problems we are currently facing, methods for robust detection of specific RNA in cells are more important than ever. While a few technologies that link RNA signals to biomolecular outputs have emerged, the field has yet to converge on a tool that is easily customized to different endogenous RNAs and that can be paired with virtually any biomolecular output of choice. To address these issues, we have developed a novel plug-and-play RNA detection platform. This platform uses synthetic split ribozymes which, in the presence of a user-specified RNA target input, produces an mRNA output that is translated into protein. Because sensing is achieved through designable base-pairing interactions between target RNA and guide RNAs that are appended onto the split ribozyme, redesign of sensors for new targets is simple and straightforward. To deliver this system, I first identified the optimal ribozyme split site that yields a high-dynamic range of sensing, where we observed no detectable output protein expression in the absence of input RNA. I then established guide RNA design rules that allow for more predictable design of new sensors, and developed a modular approach whereby any protein output can be rapidly exchanged for another without system re-engineering. Finally, I demonstrated the ability of this system to sense cellular RNAs, thereby creating sensors for dynamic detection of cell state and phenotype. Future work will focus on improving generalizability by implementing this platform in different cell types and using it for diagnostic and therapeutic applications in biotechnology.

Engineering a Protein Biomaterial: 2D and 3D Cell Culture Applications of Caf1

Peters, D. T. - Presenter, Newcastle University

Waller, H., Newcastle University

Creigh, A., Newcastle University

Dura, G., Newcastle University

Gulati, P., Newcastle University

Birch, M. A., Cambridge University

Reynolds, N. J., Newcastle University

Chrzanowska-Lightowlers, Z. M., Newcastle University

Lightowlers, R. N., Newcastle University

Fulton, D. A., Newcastle University

Lakey, J. H., Newcastle University

The environment of mammalian cells is as important *in vitro* as in the body, allowing their growth and adherence, and shaping their behaviour. In their natural environment, the extracellular matrix (ECM) contains all the signals, mechanical and biochemical, which the cells need in order to function correctly, and the challenge of any biomaterial is to recreate these signals effectively.

The choice of available materials is vast, but in most cases is complicated by either the complexity of conjugating bioactive signals to an inert material, or else using a highly bioactive but poorly defined material that is often derived from an animal. Protein materials have received little attention on account of their problems with low stability and a tendency to aggregate, however by using protein engineering tools we can overcome these disadvantages, and instead benefit from the flexibility that protein structure and chemistry can deliver.

In this work, we demonstrate how the bacterial protein, Caf1 - which is non-bioactive, highly stable and free from animal material â€" can be engineered to support the growth of various cell types, including induced pluripotent stem cells (iPSCs). We then show how Caf1 can be used in the differentiation of stem cells down both heart and bone lineages, and how it may function in a wound healing context. Finally, we describe how the innate capacity of Caf1 to refold and self-assemble following thermal denaturation allows both increased biological complexity and new material possibilities, providing an easy route for 3D cell culture.

Engineering Alternative Degradation Tags for Synthetic Circuits

Jadhav, P. - Presenter, South Dakota State University

Butzin, N. C., Virginia Tech

The goal in synthetic biology is to build robust synthetic circuits in bacteria that are dynamic, highly regulated, and result in a unified response. In the last 20 years, the synthetic biology field has effectively leveraged transcriptional (RNA production) and translational (protein production) controls. However, protein degradation plays an important role in determining the half-life of proteins and regulating biological systems. Amino acid degradation tags are exploited to avoid a reliance on cell division for protein dilution and to build dynamic circuits. The most leveraged degradation tag in *E. coli* is the ssrA-tag, which is an 11-amino acid sequence that primarily target proteins for degradation by the ClpXP proteolytic system. However, the use of the ssrA-tag limits scalability and complexity especially in synthetic oscillators in bacteria due to its multiple proteolytic target recognition signals. Our goal is to build orthogonal oscillators that utilize degradation tags targeting to multiple proteases with no crosstalk. In this study, we are modifying the ssrA-tag to change the substrate affinity and degradation rate and produce new synthetic oscillators. We hypothesize that changing the degradation rate can alter the output signal of the oscillator. We have tailored the ssrA-tag to reduce crosstalk between proteolytic systems to increase robustness and developed a variety of degradation tags. While many of these tags exhibited decreased or no change in degradation rates, one depicted a significant increase. We further tested interesting candidates for crosstalk between proteolytic systems and identified a tag that display minimum to no crosstalk with other proteolytic systems. We aim to build and compare the output signals of different oscillators with novel tags in batch cultures and at the single-cell level. The design and implementation of these novel degradation tags will enable development of biological building blocks for increased complexity in synthetic circuits.

Engineering and Directed Evolution of Cellular Protein Translation

Badran, A. - Presenter, The Broad Institute of MIT & Harvard

Rapid methods to engineer evolve and cellular translation machinery would enable applications beyond canonical protein synthesis. Unsurprisingly, manipulations to cellular translation often yield pleiotropic results, confounding experimental interpretation as researcher-dictated activities cannot be easily uncoupled from cellular fitness. Here, we develop a series of systematic advances to engineer and evolve novel translational capabilities in living cells. We apply orthogonal translation to define general requirements for efficient heterologous rRNA processing. We discover that supplementation with a small subset of cognate r-proteins enhances heterologous ribosome activity for rRNAs derived from organisms with as little as 76.1% 16S rRNA sequence identity to *E. coli*. We leverage these advances to develop oRibo-PACE, a novel method for continuous evolution of orthogonal rRNAs. Using oRibo-PACE, we evolved orthogonal ribosomes from *E. coli*, *P. aeruginosa* and *V. cholerae* for >260 generations under increasing selection regimes. Evolved o-ribosomes exhibited improved protein translation activities, up to 400% of corresponding wild-type ribosomes, at similar or reduced cellular burdens. Finally, we develop an engineering approach that affords quadruplet-decoding tRNAs using representative scaffolds for all 20 canonical amino acids, and employ novel PACE methodologies to substantially improve their translation efficiency without compromising amino acid selectivity.Collectively, our results simplify translational engineering and directed evolution pipelines, and provide extensible strategies for affecting ribosome function *in vivo.*

Engineering Bacteroides Thetaiotaomicron to Secrete Heterologous Therapeutic Protein

Lyu, S. - Presenter, University of Illinois at Urbana Champaign

Protein drugs, including engineered antibodies, present many advantages, including high specificity, high potency, and high biocompatibility. To prevent degradation in the stomach, past studies have attempted to engineer probiotic bacteria to produce protein drugs in situ, within the gastrointestinal tract. *Lactococcus lactis* has been the most commonly used species for this purpose because of its â€œgenerally recognized as safeâ€ (GRAS) status and the availability of tools for genetic manipulation of this organism. However, *L. lactis* only transiently colonizes the human gut and therefore requires repetitive dosing to maintain a therapeutic effect. Our research aims to engineer a prevalent and abundant permanently colonizing gut commensal bacterium, *Bacteroides thetaiotaomicron*, to secrete small antibody variants for the treatment of intestinal diseases. To facilitate engineering of *B. thetaiotaomicron*, we created a Golden Gate Assembly system to enable rapid construction of plasmids with a variety of genetic parts. We identified eleven signal peptides by bioinformatic analysis and three therapeutic antibodies against various targets of gastrointestinal disease (two targeting *Clostridioides difficile* infection and one targeting chronic colitis), which we have engineered into smaller variants to accommodate bacterial expression and increase stability in the GI tract. To screen for secretion activity, we will use our plasmid assembly system to pair each of the signal peptides with each of the different antibody variants, and then measure antibody accumulation in culture media. Preliminary results have revealed two signal peptides supporting 15-50 times higher levels of antibody secretion from *B. thetaiotaomicron* than previously reported in another *Bacteroides speices.* Future work will include engineering an improved secretory pathway to optimize secretion. Ultimately, we aim to establish engineered *B. thetaiotaomicron* strains as self-replicating drug production factories for *in situ* delivery of active biotherapeutic proteins with the goal of developing a platform for single-dose, low-cost, long-term therapy for chronic intestinal diseases.

Engineering Endosymbionts Capable of Controlling Mammalian Cell Gene Expression

Madsen, C. S. - Presenter, Michigan State University

Makela, A. V., Michigan State University

Hardy, J. W., Michigan State University

Contag, C. H., Stanford University

Directed cellular reprogramming has the potential to impact human health through guiding tissue regeneration, advancing cell-based therapeutics and modulating immune responses. However, controlling cellular fates and function in complex tissue environments has remained a significant challenge. To address this challenge, we have developed an engineered endosymbiont (EES) designed to escape destruction in mammalian cell vacuoles and reside in the cytoplasm where it can control gene expression and guide host cell fate and function. EES can be engineered with regulatable operons that can be induced to express and secrete reporter genes for imaging, or transcription factors that direct cellular function. We evaluated EES for modulating immune cells both as a demonstration of EES-driven cellular programming, and as a means of controlling the host immune response for therapeutic applications. After transfer to host cells, the EES was shown to escape endosomes and replicate within the cytoplasm via confocal and live cell microscopy. We used the *lacZ* gene encoding Î^2-galactosidase with a secretion sequence and nuclear localization sequence to demonstrate delivery of proteins made in the EES to the nuclei of host cells as visualized by fluorescence microcopy. Operons encoding transcription factors, *Stat-1* and *Klf6* or *Klf4* and *Gata-3,* capable of polarizing macrophages into M1 or M2 phenotypes, respectively, were developed and cloned into the EES. Intracellular EES may be able to polarize resting state J774 macrophage/monocyte cells into M1 or M2 phenotypes after induction of transcription factors. Accordingly, the EES could be used as a therapeutic for treating cancer or arthritis by polarizing macrophages to M1 or M2 phenotype, respectively. EES are versatile cellular control modules that could direct cellular functions for the treatment of diseases, or for guided regeneration and replacement of lost or damaged tissues. The regulation of immune function would demonstrate the utility of EES to control regenerative processes.

Engineering Flavin Synthesis in Bioelectronic E.coli for Multichannel Sensing

Zhang, X. - Presenter, Rice university

Ajo-Franklin, C., Lawrence Berkeley National Laboratory

Microbes can exchange the electron with their external environment, such as iron and electrode. These extracellular electron transfer processes can be developed into bioelectronic devices, which can convert biochemical information into electrical response. One such example for a bioelectronic sensor is Mtr-expressing *E.coli*, which incorporated the Mtr pathway from *Shewanella oneidensis* MRâ€1 to report sensing signal as current.[1,2] However, sensing signal in Mtr *E.coli* is limited to only one output channel. To provide a second reporting channel, we have introduced flavin, which was demonstrated as a redox mediator with different redox potential than Mtr pathway, to have two distinguished redox channels. By engineering the riboflavin pathway from *Bacillus subtilis* into Mtr-expressing *E.coli*, we were able to double the extracellular flavin concentration than the single Mtr *E.coli*. With more flavin production, the signal output as current increased from flavin-Mtr expressing *E.coli.* We can also distinguish different redox peaks from flavin-Mtr *E.coli* via electrochemical approaches. To further build up bioelectronics based on flavin-Mtr *E.coli*, an hydrogel-encapsulated system was utilized as a bioelectronic-capsule. These enhancements open new possibilities for engineering more sophisticated and sensitive techniques, such as multi-channel biosensors.

1. TerAvest, M. A., Zajdel, T. J. & Ajoâ€Franklin, C. M. The Mtr Pathway of Shewanella oneidensis MRâ€1 Couples Substrate Utilization to Current Production in Escherichia coli. *ChemElectroChem* (2014).

2. Su, L., Fukushima, T. & Ajo-Franklin, C. M. A hybrid cyt c maturation system enhances the bioelectrical performance of engineered Escherichia coli by improving the rate-limiting step. *Biosens. Bioelectron.* **165**, 112312 (2020).

Engineering of Rare Earth Binding Proteins for Greener Extractions

Reinert, Z. E. - Presenter, Air Force Research Laboratory

Hung, C. S., Air Force Research Laboratory

Poole, A. R., Air Force Research Laboratory

Perminova, D., UES, Inc.

Gomez-Santos, K., UES, Inc.

Hinkle, K., University of Dayton

Berry, R., USAF

Rare earth elements (REEs) are a group of metals with a vast range of applications including electric motors, lasers, GPS, and batteries. Unfortunately, extraction and separation of REEs has considerable negative environmental impacts due to the use of large amounts of acidic and organic solvents. Separation of REEs is further complicated by the highly similar chemical and physical properties of REEs and their relatively low abundance compared to other metals such as calcium and magnesium. To date, no synthetic molecules have demonstrated selectivity within REEs. Recently, a protein known as LanModulin (LanM) was discovered that has markedly high affinity for REEs, even in the presence of millimolar concentrations of calcium. We hypothesized that LanM and other metal-binding proteins could serve as a starting point to engineer proteins with high selectivity for select REEs. We have begun dissecting the fundamental forces of REE-binding in LanM using computational and experimental efforts. The metal binding affinity of individual domains of LanM has been investigated via UV/Vis spectroscopy. We will leverage our results to design improved variants with high selectivity for specific REEs. These LanM analogues will be deployed to develop greener methods for REE extraction and allow for new sources of REEs (such as electronic waste) to be utilized.

Engineering Protein Secretion Platforms for Living Materials in Caulobacter Crescentus

Orozco Hidalgo, M. T. - Presenter, LBNL

Living organisms create biomaterials with exceptional mechanical properties. These biomaterials also respond to various environmental cues and can self-generate & self-heal [1]. These characteristics lead to the construction of the engineered counterparts: engineered living materials (ELMs). However, our ability to create these new biomaterials is largely limited by how well we can program living organisms to synthesize biomolecules at the right place and amount [1]. Here we programmed *Caulobacter crescentus* to secrete proteins via the Type I Secretion System (T1SS) and the Type II Secretion System (T2SS) to synthesize and functionalize biomaterials. First, we engineered the T1SS of C. crescentus to secrete an extracellular matrix protein composed of an elastin-like polypeptide (ELP) hydrogel fused to supercharged SpyCatcher [$SC^{(â˜')}$]. This biopolymer was secreted at levels of 60 mg/liter, an unprecedented level of biomaterial secretion by a native T1SS [3]. Given our success with unfolded protein secretion via the T1SS, we sought to develop our platform for folded proteins via the native T2SS. To accomplish this, we used a native lipoprotein (ElpS), substrate of T2SS [4], as a native carrier for functional folded proteins. Furthermore, to control and increase the secreted folded protein yields, we upregulated the gene expression of the T2SS apparatus via the pho regulonâ€™s response to phosphate starvation. Taken together, this work develops protein design guidelines for T1SS & T2SS in *C. crescentus* and demonstrates the autonomous secretion and assembly of extracellular protein matrices to form environmentally responsive engineered living materials.

1. Gilbert and Ellis, (2018). Biological Engineered Living Materials: Growing Functional Materials with Genetically Programmable Properties. *ACS Synth Biol.*

2. Orozco-Hidalgo *et al.*, (2021). Engineering High-Yield Biopolymer Secretion Creates an Extracellular Protein Matrix for Living Materials. *mSystems.*

3. Le Blastier *et al.*, (2010). Phosphate starvation triggers production and secretion of an extracellular lipoprotein in Caulobacter crescentus. PLoS One.

Engineering Synthetic Gene Circuits for Bacterial Therapeutic Delivery In Vivo

Harimoto, T. - Presenter, Columbia University

Chien, T., Columbia University

Hahn, J., Columbia University

Danino, T., Columbia University

Engineering bacteria to locally sense and produce therapeutics in disease sites such as tumors is emerging as a promising application of synthetic biology. However, a central challenge for translating this next-generation therapy to clinics is the precise control of bacterial behavior *in vivo*. We recently constructed a library of biosensors and therapeutic payloads, and rapidly characterized their performance in a novel 3D multicellular co-culture platform that recapitulates bacterial tumor colonization. Using this system, bacteria were engineered to selectively grow in tumor-specific environmental conditions and deliver payloads. In murine cancer models, we demonstrated that the engineered bacteria improved safety and efficacy profiles. These approaches can be utilized towards accelerating the development of therapeutic bacteria for clinical applications.

Equalizer Circuits Provide Constant Gene Expression Regardless of Plasmid Copy Number in Mammalian Cells

Lee, J. - Presenter, Rice University

Yang, J., MIT

Land, M., Baylor College of Medicine

Lai, S., Baylor College of Medicine

Igoshin, O., Rice University

Precise control of gene expression is critical for biological research and biotechnology. However, transient plasmid transfections in mammalian cells produce a wide distribution of plasmid copy numbers per cell, and consequently, high expression heterogeneity. High expression heterogeneity can make genetic circuits less predictable and unreliable. Here, we report plasmid-based synthetic circuits â€" Equalizers â€" that buffer copy number variation at the single-cell level. Equalizers couple a transcriptional negative feedback loop with post-transcriptional incoherent feedforward control. We demonstrate that Equalizers outperform other gene dosage compensation topologies and produce as low cell-to-cell variation as chromosomally integrated genes. We also show that episome-encoded Equalizers enable a rapid generation of extrachromosomal cell lines with stable and uniform expression for over two months. Overall, Equalizers are simple and versatile devices for homogeneous gene expression and can facilitate the engineering of synthetic circuits that function reliably in individual cells.

Evolutionary Innovation By Hypermutation Using a Targeted Artificial DNA Replisome

Yi, X. - Presenter

Khey, J., University of Minnesota

Kazlauskas, R. J., University of Minnesota

Travisano, M.

Extensive exploration of a protein's sequence space for improved or new molecular functions requires *in vivo* evolution with large populations. But disentangling the evolution of a target protein from the rest of the proteome is challenging. Here we designed a protein complex of Targeted Artificial DNA Replisome (TADR) in live cells to processively replicate one strand of a plasmid with errors. It enhanced mutation rates of target plasmid up to 2.3Ã—10^{5}-fold with only a 78-fold increase in off-target mutagenesis. It was used to evolve itself to increase error rate and increase the efficiency of an efflux pump while simultaneously expanding the substrate repertoire. TADR enables multiple simultaneous substitutions to discover functions inaccessible by accumulating single substitutions, affording potential for solving hard problems in molecular evolution and development of biologic drugs and industrial catalysts.

Evolution-Guided Engineering of Elongation Factor-Tu for Incorporation of Noncanonical Amino Acids

Authors:

Xiong, L. - Presenter, Georgia State University

Martinez Bello, L., Georgia State University

Matsuura, M. F., The Westheimer Institute for Science and Technology

DeLey Cox, V. E., Georgia Institute of Technology

Gaucher, E. A., Georgia State University

Site-specific incorporation of noncanonical amino acids (ncAA) has emerged as a promising tool for synthesizing proteins with novel biological properties. However, the breadth of ncAA incorporation systems is limited by incompatibility between ncAA and endogenous translational machinery. Specifically, Elongation Factor Tu (EF-Tu), a critical translational component that proofreads and delivers the aminoacyl-tRNAs to the ribosome, has been a major barrier for ncAA incorporation. Here we employ reconstructing evolutionary adaptive paths (REAP) analysis and create a site-saturation library based on 48 positions of EF-Tu that potentially modulate its recognition of ncAA-tRNA complexes. Using *in vivo* antibiotic resistance-based selection assays and fluorescence-based screening assays, we isolate 18 EF-Tu variants that could incorporate either *O*-phospho-L-serine or 4-azido-L-phenylalanine via an amber codon. To further evaluate EF-Tu variants that have shown substrate-promiscuity *in vivo*, we construct an *in vitro* assessment system based on the PURExpress system (New England Biolabs). We then examine EF-Tu variantsâ€™ abilities to incorporate exotic ncAAs that are not compatible with aminoacyl-tRNA synthetases with the help of flexizymes. These *in vitro* assays allow us to interrogate EF-Tu variants for their individual abilities to incorporate ncAAs. Collectively, these findings support an evolutionary approach to identify poly-specific or ncAA-specific EF-Tus and broaden the scope of ncAA incorporation.

Expanding Nature's Catalytic Toolbox with Artificial Metalloenzymes

Vornholt, T. - Presenter, ETH Zurich

Jeschek, M., ETH Zurich

Panke, S., ETH Zurich

Ward, T. R., University of Basel

Nature offers an immense variety of enzymes that we can employ to construct metabolic pathways for the production of compounds of interest. Nonetheless, to realize the full potential of synthetic biology, we ultimately need to move beyond the natural repertoire of enzymes and reactions. Artificial metalloenzymes (ArMs), which harness the catalytic potential of transition metals to catalyze new-to-nature reactions, offer exciting opportunities in this regard. However, most ArMs reported to date are only moderately active and barely function in living cells. To change this, we have established a whole-cell screening platform for ArMs that relies on periplasmic catalysis in *Escherichia coli*. Combined with lab automation and machine learning, this allows us to rapidly and efficiently optimize ArMs towards increased activity *in vivo*. Using this platform and a concise library of 400 double mutants, we recently identified improved ArMs for five bioorthogonal reactions, which are catalyzed by enzymes that harbor gold and ruthenium cofactors. The obtained scaffolds represent valuable starting points for larger, machine learning-guided engineering campaigns, which are currently under way to push the boundaries of ArM catalysis *in vivo*.

Our ultimate goal is to enable new-to-nature transition metal catalysis in cells in order to construct novel metabolic pathways. To achieve this, we further engineer *E. coli* chassis that address key challenges in the field, featuring improved uptake of artificial cofactors and mutual compatibility between cell physiology and bioorthogonal catalysis. To this end, we combine microfluidics and fluorescence-activated cell sorting to facilitate genome-wide knockout and overexpression studies with the goal of identifying and engineering targets that influence ArM activity *in vivo*.

We expect that these combined efforts to systematically engineer both ArMs and host organisms will enable various new applications in fields such as biocatalysis, metabolic engineering and xenobiology.

References:

Vornholt, Jeschek, *ChemBioChem*, 2020

Vornholt et al., *Science Advances*, 2021

Exploiting Microbial Cell Factories for the Development and Manufacturing of Novel Biologics and Industrial Products

Cruz, R. - Presenter, Ingenza Ltd

Establishing bioprocesses to manufacture valuable biologics and industrial products in microrganisms requires the design, construction and optimisation of strain productivity and stability, fermentation and efficient DSP. This presentation will exemplify Ingenzaâ€™s integrated toolbox of enabling technologies in industrial biotechnology and engineering biology. Ingenzaâ€™s combinatorial genetics (inABLEÂ®), genome editing, advanced HTP screening methods, FACS-assisted optimization of protein expression and tailored fermentation protocols combine to rapidly demonstrate bioprocess feasibility to meet manufacturing requirements. Case studies will be focused on directed evolution of novel enzymes, cost-competitive recombinant production of a first-in-class antibacterial peptide and establishment of a plug-and-play microbial vaccine platform meeting future global needs. Our bioengineering technologies are bringing increased predictability to overcome persistent limitations associated with strain improvement, enabling rapid progression to process optimization and scale-up for applications.

Fabio: Fatty Acid Biosynthesis: Inducible, Orthogonal

Goyal, G. - Presenter, Joint BioEnergy Institute

Engineering heterologous fatty acid metabolic pathways (e.g., for the production of biofuels and other bioproducts) in E. coli (and in other organisms) is complicated and made much more difficult by concurrent and highly regulated endogenous fatty acid metabolism (and catabolism). Since fatty acid metabolism is essential for cellular growth, and the genes responsible for it are distributed across the genome, it is not possible to simply remove it wholesale; nor is it straightforward how to repress it in a controllable fashion. Our proposed approach is to remove from the genome (via CRISPR Cas9 approach) the genes responsible for endogenous fatty acid metabolism and place them (under native gene regulation) instead on a replicating artificial chromosome that can be controllably degraded (via CRISPR Cas9 and then subsequent endogenous nucleases).The heterologous fatty acid pathway (either plasmid borne or chromosomally-integrated) could then be induced in a manner orthogonal to and independent from the endogenous metabolism. Our main goal in this research is to decouple fatty acid biosynthesis in this system from, cell growth, self-regulation, consumption of fatty acids for other biomolecules such as phospholipids and lipopolysaccharides.

Fabrication and Cross-Chassis-Characterisation of Artificial Promoters and 5′ Untranslated Regions in Vibrio Natriegens and Escherichia coli

Tietze, L. - Presenter, NTNU

Vibrio natriegens is a rising star on the horizon of microbial gene expression chassis, growing even faster than model organism *Escherichia coli*. To express genes in *Vibrio natriegens*, we currently use genetic elements optimized for *Escherichia coli* because few of *Vibrio natriegens*â€™ native genetic elements are known and characterized. This means that *Vibrio natriegens*â€™ full potential as a gene expression chassis remains unknown. Characterizing native genetic elements is a cost- and time-consuming process. To reduce time and cost, we used the Gene Expression Engineering (GeneEE) method to efficiently create and characterize artificial genetic elements (promoters and 5Â´ Untranslated Regions) that lead to expression of Green Fluorescent Protein and mCherry in a wide range of expression levels. We characterized the artificial genetic elements in *Vibrio natriegens* and *Escherichia coli* and found that some of the artificial genetic elements lead to high expression in *Vibrio natriegens* but not in *Escherichia coli*. With our study we expand the pool of known and characterized genetic elements in *Vibrio natriegens*, which can be used for many different biotechnological applications utilizing *Vibrio natriegens* as a gene expression chassis.

FireProtASR: Fully Automated Pipeline for Resurrecting Stable and Expressible Ancestral Proteins

Khan, R. T. - Presenter, Loschmidt Laboratories

Musil, M., Loschmidt Laboratories

Stourac, J., Loschmidt Laboratories

Bednar, D., Loschmidt Laboratories

Damborsky, J., Masaryk University

Robust and stable enzymes accelerating chemical reactions with high specificity represent one of the keystones of metabolic engineering. During the course of evolution, nature produced a large number of diverse solutions to enzymatic catalysis by evolving natural proteins. While most of these ancestral proteins have been lost due to extinction, or to the march of evolution, it is possible to statistically infer their sequences using Ancestral Sequence Reconstruction (ASR)[1]. ASR approach has been used to resurrect ancestral proteins, to identify key amino acid residues in metabolic enzyme complexes[2], to engineer thermostable enzymes[3], to deduce information about evolutionary events and natural history[4], as well as, to engineer enzymes with dual, catalytically distinct activities[5]. The latter is possible by reconstructing ancestors of enzymes that have catalytically differentiated, and evolved along separate trajectories.

ASR is generally not accessible to scientists outside communities of evolutionary biologists. The purpose of this research is to construct a fully automated pipeline that allows anyone, including those who lack specialist knowledge, to perform ASR, thus reducing the academic barrier to entry. The pipeline was verified against work that was previously done in the lab; on the ancestral sequence reconstruction of haloalkane dehalogenase family[5]. The pipeline is accessible online through the FireProt^{ASR} web server [6] at: https://loschmidt.chemi.muni.cz/fireprotasr/. The server can be used for designing stable, highly expressible and catalytically active enzymes for metabolic pathways, that can be used for the construction of cell factories producing high-value chemicals.

REFERENCES

1. Pauling, L., & Zuckerkandl, E., *Acta Chemica Scandinavica Supplements* **1963**, 17, S9â€"S16.
2. Holinski, A., et al., *Proteins: Structure, Function, and Bioinformatics* **2017**, 85.2, 312-321.
3. Babkova, P., et al., *ChemBioChem* **2017**, 18, 1448.
4. Gaucher, E., et al., *Nature* **2008**, 451, 704
5. Musil, M., et al., *Briefings in Bioinformatics* **2020**, bbaa337, 1.
6. 6. Khan, R. T., et al., *Current Protocols* **2021,** 1.2, e30.

Glycosylated Biofilm Proteins for Functional Biomaterials

Sahin Kehribar, E. - Presenter, Bilkent University

Isilak, M. E., Bilkent University

Bozkurt, E. U., Bilkent University UNAM

Seker, U., Bilkent University

Adamcik, J., ETH Zurich, Department of Health Sciences and Technology

Mezzenga, R., ETH Zurich, Department of Health Sciences and Technology

Amyloids are fibrillar protein aggregates, dominated by Î²-sheets in their structure. Historically, amyloids are associated with diseases, however several functional amyloids are also identified. Biofilm proteins from microorganisms such as CsgA of *Escherichia coli* or TasA of *Bacillus subtilis* are great examples of functional amyloids with numerous applications as functional biomaterials. Due to their high mechanical strength, rigidity and resistance to harsh environmental conditions, amyloids emerged as outstanding biomaterials.

Protein glycosylation is one of the most common post-translational modifications, which is classified into two groups as N-linked and O- linked glycosylation. Protein glycosylation is firstly identified in eukaryotic cells; however, bacterial glycosylation is also discovered in the past twenty years, especially associated with pathogenicity. The first identified prokaryotic species that possess N-linked glycosylation is *Campylobacter jejuni*, a pathogenic bacterium that causesgastroenteritis. The *pgl* glycosylation pathway from *C. jejuni* is transferred to *E. coli* and successful protein glycosylation is achieved. Glycosylation is crucial for adhesiveness of proteins. Therefore, glycosylation can be utilized as a practical tool to alter adhesive properties of target proteins for protein-based biomaterial applications.

Given the fact that amyloid-like TasA fibrils are utilized in biomaterial research, we hypothesized that adhesiveness of TasA fibrils may be enhanced via glycosylation. To that end, TasA is engineered to have DQNAT glycosylation motif at the C terminal and expressed it in *E. coli* containing *pgl* circuit. Adhesive and viscoelastic properties of TasA fibrils are examined on solid surfaces. We showed that glycosylation can indeed enhance adhesive properties of TasA fibrils without disrupting fibrillization. Our results demonstrated that *de novo* glycosylation can be used as a tool to produce biomaterials with superior adhesive performance [1]. Supported by TUBITAK 216M127.

[1] E. Sahin Kehribar, M. E. Isilak, E. U. Bozkurt, J. Adamcik, R. Mezzenga, U. O. S. Seker, *Biomater. Sci.* **2021**, DOI 10.1039/D0BM02192J.

gRNA-Seqret: Genome Wide Guide RNA Design and Sequence Extraction Tool

Simirenko, L. - Presenter, DOE Joint Genome Institute, Lawrence Berkeley National Laboratory

Blaby, I. K., The Joint Genome Institute

Cheng, J. F., Lawrence Berkeley National Laboratory

The U.S. Department of Energy (DOE) Joint Genome Institute (JGI) provides DNA sequencing and synthesis services to scientific users via community science programs. The DNA synthesis program covers all aspects of the synthetic biology design-build-test cycle, enabling users to engineer biological systems of relevance to their research and the DOE mission.

One particular product type offered by the DNA synthesis program is the design and construction of libraries with high degrees of variants. Applications of such libraries include CRISPR technologies using guide RNA (gRNA) libraries or studying structural variations of expressed proteins or regulatory elements (e.g., promoters).

Presently, there are no simple tools for extracting sequences from specific genomic regions in bulk, and most available tools for gRNA design restrict the user to pre-processed genomes from specific model organisms. To overcome these shortcomings, we have developed a web-based open-source *in silico* design tool: The guide RNA and Sequence Region Extraction Tool (gRNA-SeqRET). This tool allows the user to specify exactly which region of a gene, a subset of genes, or every gene in the genome for sequence extraction; a number of criteria can be used to specify gRNA targets.

The initial release of the software allows the extraction of specific sequences upstream and downstream of gene start sites and whole genome gRNA libraries for both inhibiting (â€œCRISPRiâ€) and activating (â€œCRISPRaâ€) genes in any prokaryotic organism. Furthermore, it evaluates each gRNA based on predicted RNA folding with and without the scaffold sequence, and automatically selects the top candidates. Further development in progress will extend the toolsâ€™ capabilities to use a wider variety of genome sequence and annotation file formats, the capability to process eukaryotic genomes, and the addition of more flexible sequence extraction and targeting options decoupled from the gRNA design workflow, providing a universal sequence extraction tool.

Identifying Yeast Gene Deletions That Enhance Noncanonical Amino Acid Incorporation

Zackin, M. T. - Presenter, Tufts University

Stieglitz, J. T., Tufts University

Van Deventer, J., Tufts University

Efficient incorporation of noncanonical amino acids (ncAAs) in proteins is a prerequisite for leveraging their chemical diversity in the engineering of proteins with expanded capabilities. Potential applications for ncAAs span from engineering biological therapeutics with chemistries beyond those found in canonical amino acids to expanding our fundamental understanding of protein biosynthesis. Efforts to improve ncAA incorporation have focused on engineering orthogonal translation systems (OTSs): the translation machinery introduced into cells that allows for the encoding of ncAAs in proteins. However, the efficacy of these efforts is limited by complex cellular factors that inhibit genetic encoding of ncAAs at repurposed stop codons. Identification of these cellular factors is necessary to engineer cells that better accommodate expanded genetic codes. In this work, we performed a genome-wide screen of *Saccharomyces cerevisiae* using a pooled yeast knockout (YKO) collection to identify single-gene deletions that enhance ncAA incorporation efficiency and fidelity. We utilized dual-fluorescent reporter systems that facilitated fluorescence-activated cell sorting (FACS) of the YKO collection to isolate deletion strains with increased ncAA incorporation efficiency. Our early screens identified several unique deletions that appear to exhibit improved stop codon suppression and ncAA incorporation. Identified genes are involved in diverse cellular activities, and genes with no previously characterized function have also been identified. Current efforts focus on further validating these deletions, including verifying that these single-gene deletions enhance ncAA incorporation efficiency in other strains, and quantitatively evaluating ncAA incorporation efficiency and fidelity. Future work will interrogate these single-gene deletion strains to elucidate how they affect ncAA incorporation for applications such as click chemistry. To our knowledge, this work is the first genome-wide investigation of factors that influence ncAA incorporation. Understanding the cellular processes that dictate stop codon suppression and ncAA incorporation are imperative to augmenting genetic code manipulation strategies for a multitude of applications.

In Vitro Fermentation Test Bed for Evaluation of Engineered Sense and Respond Probiotics in Polymicrobial Communities

Arcidiacono, S. - Presenter, CCDC - Soldier Center

Doherty, L. A., U.S. Army Natick Soldier Research Development and Engineering Center

Whitman, J., NSRDEC

Pantoja-Feliciano, I., U.S. Army Natick Soldier Research Development and Engineering Center

Soares, J., Natik Solider Research, Development and Engineering Center

Goodson, M. S., Air Force Research Laboratory

Breedon, A. M. E., UES, Inc.

Lyon, W., US Air Force Research Laboratory

Jimenez, G., Air Force Research Laboratory

Walper, S., US Naval Research Laboratory

Spangler, J. R., National Research Council

There is great interest in using synthetic biology to engineer organisms with novel or improved function to add to the human gut microbial community; including engineering probiotics for increased beneficial function. The design-test-build-learn approach is commonly used in synthetic biology, largely for circuit design. Engineered bacterial cells are developed and tested in monoculture, under optimal growth conditions that donâ€™t reflect the highly complex environments of the intended application. In contrast to monoculture, characterizing engineered organism behavior within a highly complex microbial community can be challenging. *In vitro* fermentation models using simplified microbial communities address this challenge by reducing complexity and allowing the study of organism persistence and function within more tailored microbial community dynamics. With increased experimental capacity, *in vitro* models allow rapid evaluation of multiple parameters to generate knowledge that could then inform design of more physiologically relevant *in vivo* studies. Here, simplified polymicrobial communities of gut organisms were utilized to investigate engineered probiotic bacteria *Escherichia coli* Nissle and *Lactobacillus plantarum*. A simplified community with increasing complexity influenced reporter output with multiple plasmid constructs that encode for varying fluorophores. The function of the engineered *E. coli* Nissle constructs, designed to produce GFP, was dependent on community composition. *L. plantarum* engineered to produce GFP showed comparable functional effects. The simplified community of gut bacteria was chosen based on not only abundance on substrate utilization and metabolism. Engineered organism function and persistence tested in the presence of simplified communities of increasing complexity to determine: 1) engineered organism persistence and; 2) effect of engineering on organism persistence; and 3) engineered organism effect on the commensal composition. Testing engineered probiotics using an *in vitro* fermentation test bed approach can provide critical knowledge for circuit design feedback and functional validation prior to higher fidelity testing within animals or humans.

Interpretable Modeling of Genotype-Phenotype Landscapes without Sacrificing Predictive Power

Tonner, P. D. - Presenter, NIST

Pressman, A., NIST

Tack, D. S., University of Texas at Austin

Ross, D., National Institute of Standards and Technology

Large-scale measurements linking genetic sequence (genotype) to corresponding function (phenotype) have grown steadily in size and scale, with genotype-phenotype landscape (GPL) datasets typically containing 10^5 observations or more and assaying thousands of genetic mutations. Despite this growth, the space of all possible genotypes is larger than can ever be measured. Predictive modeling is necessary to address this gap. Additionally, our ability to predict genotype-phenotype relationships directly impacts the speed and accuracy with which we can bioengineer organisms, study the evolution of pathogens, and evaluate molecular biomedicine.

Black-box neural networks are currently the dominant choice of GPL models due to their unsurpassed ability to generate accurate out-of-sample predictions. These models suffer, however, from their inherent inability to explain their predictions. While methods for post-hoc explanation of neural network predictions exist, they can only approximate the trained model, and the accuracy of those approximations is often difficult to assess. Despite these issues, neural networks remain popular due to the assumption that there is a necessary trade-off between a model's predictive accuracy and its interpretability.

As an alternative, we developed an approach to modeling GPLs that is inherently interpretable, called LANTERN. LANTERN learns a low-dimensional latent phenotype space where mutations combine additively. The latent phenotype is then transformed to observed phenotype measurements through a smooth, non-linear function. This approach ensures that the predictions can be easily decomposed into interpretable components. Despite this simplicity, LANTERN provides predictive accuracy equal to or better than neural network methods when applied to GPL data.

LANTERN also provides novel metrics of GPL structure, including the empirical dimensionality, local additivity, and phenotypic robustness. Notably, these are learned *de-novo* from data, without additional domain specific knowledge. Overall, LANTERN demonstrates that there is no necessary tradeoff between predictive accuracy and interpretability for GPL data.

Listen to DAD (Data-optimized Assembly Design) When Constructing High-Complexity Golden Gate Assembly Targets

Pryor, J. M. - Presenter, New England Biolabs

Potapov, V., New England Biolabs, Inc.

Pokhrel, N., New England Biolabs

Lohman, G. J. S., New England BioLabs, Inc.

Large DNA constructs (>10 kb), including small genomes and artificial chromosomes, are invaluable tools for genetic engineering and vaccine development. However, the manufacture of these constructs is laborious. To address this problem, we applied new design insights and modified protocols to Golden Gate assembly. While this methodology is routinely used to assemble 5-10 DNA parts in one-step, we found that optimized assembly permitted >50 DNA fragments to be faithfully assembled in a single reaction. We incorporated these findings into a suite of webtools that design assembly reactions, and applied these insights to genome construction, carrying out rapid assembly of the 40 kb T7 bacteriophage genome from 52 parts and recovering infectious phage particles after cellular transformation. The new Golden Gate assembly webtools, protocols, and design principles can be applied to rapidly engineer a wide variety of large and complex assembly targets.

Living Biosensors at the Intersection of Genetic Circuits and Machine Learning

Bozkurt, E. U. - Presenter, Bilkent University UNAM

Saltepe, B., Bilkent University

Seker, U., Bilkent University

Güngen, M. A., Bilkent University

Çiçek, A. E., Bilkent University

Advances in synthetic biology tools pave the way for engineering cells as biosensors which take part in environmental monitoring, disease diagnostics and therapy, and bioproduction of valuable compounds. To date, remarkable successful living biosensors have been demonstrated in laboratory allowing cost-effective, user-friendly, renewable, and portable prototyping for field deployment. Yet, living biosensors may suffer from certain challenges such as leakage, specificity, sensitivity, multiplexing, and long response time. Many control elements have been designed to solve most of the challenges. However, the more control element in a circuit the more cellular burden, hence, lagging in sensor response has resulted. In the last decade, synthetic biology collaborates with other disciplines (e.g., machine learning) to solve issues regarding to circuit design. In our study, we utilized a neural network-based architecture to decrease the response time of a complex genetic circuit which is engineered for gold ion detection. Thanks to Long-Short Term-Memory (LSTM)-based networks, we were able to decrease the ON/OFF status of the sensor to 30 min with 78% accuracy and over 98% in 3 h. Next, we demonstrated that the network can also classify the sensor output (raw fluorescence data) into pre-defined gold concentration groups with 82% precision in 3 h. We envisage that this approach would be applicable to a multiple living sensors with complex gene circuits which suffers from long response time.
This study has been supported by TUBITAK Grant No 114Z653 and 118S398.

Lysate Proteome Engineering Enables High-Yield Metabolite Conversions in Crude Cell Extracts

Dinglasan, J. L. N. - Presenter, University of Tennessee

Garcia, D. C., Oak Ridge National Laboratory

Shrestha, H., Oak Ridge National Laboratory

Abraham, P. E., Oak Ridge National Laboratory

Hettich, R. L., Oak Ridge National Laboratory

Doktycz, M. J., Oak Ridge National Laboratory

Cell-free metabolic engineering (CFME) systems provide bioconversion opportunities for achieving near theoretical yields of high-value metabolites. High-yield biotransformations unattainable so far in live cells have been successfully demonstrated in purified enzyme systems. Recently, active endogenous metabolic pathways in cell extracts have been leveraged to convert simple sugars towards valuable metabolites. These lysate-based systems have several advantages over purified enzyme systems such as ease of preparation, cost, and scalability. However, due to the presence of unwanted enzymes in cell extracts, carbon flux from a supplemented substrate becomes distributed among competing pathways, potentially compromising yields of the desired end-product. To this end, we have developed a genome engineering enabled strategy to direct flux towards targeted metabolites in lysate-based cell-free systems. With the goal of concentrating carbon flux from a glucose substrate towards the central metabolic precursor pyruvate, we genome engineered *E. coli* BL21DE3 strains to express endogenous pyruvate-consuming enzymes with 6xHis-tags. These modified strains show no adverse growth phenotypes, and the proteomes of their derived extracts could be engineered. Specifically, 6xHis-tagged enzymes were removed from the derived lysates which enabled pooling of pyruvate at concentrations 40-fold higher than conventional CFME lysates. Therefore, our approach enables the reduction of flux through metabolic nodes that are difficult to engineer *in vivo* while bypassing the need to undergo multiple design-build-test cycles. This strategy was further demonstrated by an engineered lysate with an unconventional endogenous metabolic phenotype that generated 40-fold higher bioethanol than control *E. coli* lysates. Through optimization of our targeted depletion approach, as well as source strain cultivation conditions, we have further shown that this lysate can produce 0.52 g ethanol per g consumed glucose. Lysate proteome engineering can thus be used to focus flux towards targeted pathways in cell extracts, expanding the current CFME toolkit to improve bioproduction yields in lysate-based systems.

Machine Learning Guided Approach to Bacteriocin Engineering

Kay, J. - Presenter, Ingenza Ltd.

Directed evolution (DE) projects have led to many great advancements in the field of protein engineering. Commonly DE applies oversampling of the amino acid sequence space over specific regions of a protein, or else random mutagenesis of the entire sequence. In both cases, the immense potential for diversity means high throughput screening efforts are intense and often costly endeavours. Machine learning algorithms such as random forest, gradient boosted decent and support vector machine can be employed alongside traditional DE methods, to streamline development projects. These powerful algorithms model relationships between key physiochemical and structural properties and the observed protein function to identify the key design features for protein engineering. A prime application of this technology is the design of novel bacteriocins to combat bacterial antimicrobial resistance (AMR). Bacteriocins, which have bacterial origins, are broad inhibitors of many Gram-positive pathogens, such as methicillin-resistant *Staphylococcus aureus* (MRSA) and are attractive therapeutic options due to their low cytotoxicity and potential for resistance development. Much is to be learned of their exact structure-activity relationship, however it is known that engineering mutations, truncations and chimeras of two or more bacteriocins, can alter their biological function, such that their activity can be tuned or directed towards particular targets. This DE project aims to employ machine learning as a tool to study the structure-activity relationship of a family of bacteriocins, in efforts to better understand re-design these important biomolecules for use as novel antimicrobial therapies.

Macroscopic De-Novo Engineered Living Materials

Molinari, S. - Presenter, Rice University

Tesoriero, R., Rice University

Ajo-Franklin, C., Lawrence Berkeley National Laboratory

Soman, J., Rice University

Engineered living materials (ELM) are a new class of materials that contain living cells and thus have some of the unique capabilities of living systems, e.g. autonomous growth, self-repairing, or the ability to sense and respond to the environment. Most existing examples of macroscopic ELMs have introduced novel properties onto naturally-occurring biological materials. However, programming macroscopic material formation *de novo* from a model organism would offer much greater, user-defined control over material structure and function. We created a bottom-up designed ELM by engineering *Caulobacter crescentus* to display weakly interacting peptides at high density on the cell surface. This enabled the self-assembly of a novel, macroscopic ELM. We included the SpyTag peptide into the displayed protein, allowing for user-defined functionalization via SpyTag-SpyCatcher interactions. As *C. crescentus* expressing the synthetic construct grows, the cells assemble into a pellicle at the air-water interface. This layer expands in thickness and ultimately sinks into the media to form free-floating, macroscopic materials shaped as cocoons. As expected, the cocoons are predominantly composed of cells bound to each other through the synthetic protein. Surprisingly, we also observed the synthetic proteins dissociate from the cell surface to form long ~2 micron-thick filaments, which bundle into ~100-micron-thick fibers running throughout the cocoon. The morphology and size of these cocoons can be tuned by changing physical parameters, such as shaking speed and media volume. Within a cocoon, *C. crescentus* cells remain viable and can re-seed growth of new cocoons even after desiccation. Cocoons can also be processed into different 3D shapes and can be mixed with inorganic materials to form inorganic-organic composites. Taken together, this work provides design rules and a technological platform for programming bottom-up assembly of cells into hierarchically-structured, multi-functional living materials.

Modularised Engineering of Regulatory Circuits to Expand Artificial Control Mechanisms for Metabolic Engineering in Yeast

Peng, B. - Presenter, The University of Queensland

Bandari, N. C., The University of Queensland

Lu, Z., The University of Queensland

Howard, C., The University of Queensland

Scott, C., CSIRO

Trau, M., The University of Queensland

Dumsday, G., CSIRO Manufacturing

Vickers, C. E., Commonwealth Scientific and Industrial Research Organisation (CSIRO)

Precise control of heterologous pathway expression is critical to achieve optimal efficiency in microbial metabolic engineering. However, current regulatory mechanisms for transcriptional control do not deliver sufficiently precise control for many applications â€" including industrial applications â€" because they are not induced at the right time during culture growth. This is particularly problematic for the more complex regulatory control mechanisms in Eukaryotes. Predictable design of transcriptional control circuits remains a significant challenge because the common transcriptional tools are not well characterised for application in metabolic engineering. Here, we employ a modularised engineering strategy to deploy novel synthetic circuits for conditional control of heterologous pathways in *Saccharomyces cerevisiae*. To better understand operational parameters, each genetic part was characterised under industry-relevant conditions. A novel tetracycline-mediated circuit was developed to deliver stringent pathway repression to minimise metabolic burden during strain development and maintenance. This was combined with a novel temperature induction circuit to deliver an economical induction system suitable for industrial applications. This delivered a 44 % increase in production of the terpenoid nerolidol, to 2.54 g L^{-1} in flask cultivation. Using these positive/negative transcriptional regulatory circuits to control heterologous pathways expands global strategies to control metabolisms in industrial microorganisms, and provides a regulatory control mechanism which may be more suitable for industry applications.

Multi-Modal Engineering of Bst DNA Polymerase for Thermostability in Ultra-Fast Lamp Reactions

Paik, I. - Presenter, University of Texas at Austin

Ellington, A. D., University of Texas

Bhadra, S., University of Texas at Austin

Shroff, R., University of Texas at Austin

Maranhao, A. C., University of Texas at Austin

Ngo, J., University of Texas at Austin

Despite the fact that strand-displacing activity is of great utility for a variety of applications, including isothermal amplification assays, there are relatively few strand-displacing DNA polymerases. In particular, the thermotolerant DNA polymerase from *Geobacillus stearothermophilus (*previously *Bacillus stearothermophilus), Bst* DNA polymerase (*Bst* DNAP), is used in a variety of assays, including loop-mediated isothermal amplification. However, despite its wide use, its properties remain open to improvement, as has been demonstrated by a variety of engineering efforts, including the identification of point mutations that impact its robustness, strand-displacement capabilities, and nascent reverse transcriptase activity.

Engineering enzymes is typically undertaken either via high-throughput screening or directed evolution. Instead, we have brought to bear several rational design tools (domain addition, machine learning, and supercharging) to quickly and efficiently generate a vastly improved version of *Bst* DNAP, termed Br512g3, that is extremely robust and that in consequence can perform ultrafast (6â€™) isothermal amplification. It far exceeds the capabilities of its parental enzyme, Bst-LF, and a widely (almost exclusively) used engineered commercial variant, Bst 2.0, from New England Biolabs

Neutralization of Competition in Crispri Circuits Via dCas9 Feedback Regulator

Bellato, M. - Presenter, University of Pavia

Huang, H. H., MIT

Qian, Y., MIT

Pasotti, L., University of Pavia

Magni, P., University of Pavia

Del Vecchio, D., MIT

Cárdenas, P., Massachusetts Institute of Technology

CRISPR interference (CRISPRi) has emerged as a powerful tool for modulating complex endogenous transcriptional programs and for building larger scale synthetic genetic circuits [1,2].

However, highly specific sgRNA-promoter binding is, alone, insufficient to achieve independent transcriptional regulation of multiple targets. Indeed, due to competition for dCas9, the repression ability of one sgRNA reduces significantly when another sgRNA becomes expressed [3].

To solve this problem, decoupling sgRNA-mediated regulatory paths, we created a dCas9 concentration regulator that implements a negative feedback on dCas9 level [4]. This allows any sgRNA to maintain an approximately constant dose-response curve, independent of other sgRNAs.

We demonstrate the regulator performance on both single-stage and layered CRISPRi-based genetic circuits, zeroing competition effects of up to 15-fold changes in circuit I/O response encountered without the dCas9 regulator.
The efficiency of the developed regulator was confirmed in different biological contexts, varying DNA copy numbers, strains and signaling molecules activating the systems.
Our dCas9 generator is implemented in a dedicated plasmid, to be easily transported across compatible bacterial strains, thus enabling true scalability of CRISPRi-based transcriptional programs.

We establish an important new tool of broad interest for CRISPRi users to either control endogenous transcriptional programs or to create new genetic circuits for any application.

[1] Gao Y, Xiong X, Wong S, et al. Complex transcriptional modulation with orthogonal and inducible dCas9 regulators. Nat Methods 13, 12 (2016).

[2] Kiani S, Beal J, Ebrahimkhani MR, et al. CRISPR transcriptional repression devices and layered circuits in mammalian cells. Nat Methods 11, 7 (2014).

[3] Zhang S, Voigt CA. Engineered dCas9 with reduced toxicity in bacteria: implications for genetic circuit design. NAR 46, 20 (2018).

[4] Huang H-H, Bellato M, Qian Y, et al. dCas9 regulator to neutralize competition in CRISPRi circuits. Nat Commun 12, 1692 (2021).

Optical Activation of TrkB Signaling

Huang, P. - Presenter, The Chinese University of Hong Kong

Duan, L., The Chinese University of Hong Kong

Brain-derived neurotrophic factor, *via* activation of tropomyosin receptor kinase B (TrkB), plays a critical role in neuronal proliferation, differentiation, survival, and death. Dysregulation of TrkB signaling is implicated in neurodegenerative disorders and cancers. Precise activation of TrkB signaling with spatial and temporal resolution is greatly desired to study the dynamic nature of TrkB signaling and its role in related diseases. Here we develop different optogenetic approaches that use light to activate TrkB signaling. Utilizing the photosensitive protein *Arabidopsis thaliana* cryptochrome 2, the light-inducible homo-interaction of the intracellular domain of TrkB in the cytosol or on the plasma membrane is able to induce the activation of downstream MAPK/ERK and PI3K/Akt signaling as well as the neurite outgrowth of PC12 cells. Moreover, we prove that such strategies are generalizable to other optical homo-dimerizers by demonstrating the optical TrkB activation based on the light-oxygen-voltage domain of aureochrome 1 from *Vaucheria frigida*. The results open up new possibilities of many other optical platforms to activate TrkB signaling to fulfill customized needs. By comparing all the different strategies, we find that the cryptochrome 2-integrated approach to achieve light-induced cell membrane recruitment and homo-interaction of intracellular domain of TrkB is most efficient in activating TrkB signaling. The optogenetic strategies presented are promising tools to investigate brain-derived neurotrophic factor/TrkB signaling with tight spatial and temporal control.

Optimizing Nanoporeters for Multiplexable Mammalian Reporting Systems

Nguyen, A. - Presenter, University of Washington

Wilde, D. - Presenter, University of Washington

Nivala, J., University of Washington

For over 40 years, reporter systems have been used as a means to track different types of cellular activities such as gene regulation. However, the number of uniquely addressable reporters that can be used together is limited due to their readout signal overlap. For instance, simultaneous measurement of unique genetic elements with multiple fluorescent protein variants is often obscured by overlapping spectral distributions. This prevents multiplexing, which not only impacts scalability and convenience, but also the potential complexity of the system of interest. To overcome this problem, we have previously designed a new class of reporter proteins, Nanopore-addressable protein Tags Engineered as Reporters (NanoporeTERs), that are more scalable and multiplexable than traditional reporter strategies and can be read out on a commercial nanopore array. Here, we optimized this system for mammalian cell systems and characterize 9 orthogonal barcodes which can be differentiated by analyzing their nanopore current signals with machine learning. We then demonstrate this system in HEK293 cells transfected with differentially barcoded genetic circuits. Ultimately, we aim to utilize this novel reporter system to investigate complex mammalian processes such as chromatin regulation dynamics and the determination of cellular phenotypes using customized reporter cassettes.

Optogenetic Control of Antibiotic Resistance to Understand Impact of Activation Dynamics on Bacterial Survival

Sheets, M. B. - Presenter, Boston University

Dunlop, M. J., Boston University

The spread of antibiotic resistance is a major threat to human health. The single-cell dynamics surrounding the activation of resistance genes are often unknown, and insight into these dynamics can lead to improved strategies for mediating resistant infections. Specifically, the impact of resistance activation timing relative to antibiotic exposure on mixed-resistance community survival may provide insight on how to best time antibiotic doses and concentrations to prevent acquisition and proliferation of resistance.

We use an optogenetic Cre recombinase control system to regulate the activation of resistance genes over time, allowing for spatiotemporal control of resistance. Here, we characterize how the timing of resistance acquisition in a subset of the population affects growth rate and final percentage of resistant cells in the population. We present work towards understanding the impact of the resistance mechanism, initial resistant fraction, and resistance activation timing on bacterial survival.

Outline of DNA Construction Process Quality in Performance Indicator Dashboard

Meng, J. - Presenter, DOE Joint Genome Institute, Lawrence Berkeley National Laboratory

Harmon-Smith, M., Joint Genome Institute

Tarver, A., DOE Joint Genome Institute, Lawrence Berkeley National Laboratory

Evans, R., DOE Joint Genome Institute

Hillson, N. J., DOE Joint BioEnergy Institute

Blaby, I. K., The Joint Genome Institute

Cheng, J. F., Lawrence Berkeley National Laboratory

The U.S. Department of Energyâ€™s (DOE) Joint Genome Institute (JGI) DNA Synthesis Science platform has grown nearly four-fold (from 1.33 Mbp FY13 to 5.62 Mbp FY20 delivered per year). Consequently, there is increasing demand for informatics that supports DNA design and construction for increased scale and new features/tools. These informatics capabilities include SynTrack, enabling JGI scientists to submit their DNA designs for the constructions and monitor the designsâ€™ progress through the DNA assembly pipeline.

We have recently enhanced SynTrack to visualize the many years of tracked data maintained within the tool by developing a new dashboard of performance indicator charts and graphs. This dashboard functionality is built on a scalable data warehouse system and lightweight Flask server-side containers combined with Angular visualization components. The dashboard captures data from the data warehouse, generates a set of charts for analysis (e.g., DNA synthesis in bps per quarter, construction time as a function of DNA size, and construction success rates as a function of the DNA assembly method). We envision that these data visualizations and their analyses will improve DNA construction efficiency, reduce the turnaround time, and provide better process visibility. Planned future developments will allow users to view the progress of their individual project/DNA assemblies.

Phage-Assisted Evolution of Botulinum Neurotoxin Proteases with Reprogrammed Specificity

Blum, T. - Presenter, The Broad Institute of Harvard and MIT

Sequence-specific proteases have the potential to advance biotechnology and medicine, however, generation of proteases with tailor-made cleavage specificities remains a major challenge. In this work we developed a phage-assisted protease evolution system with simultaneous positive and negative selection and applied it to three botulinum neurotoxin (BoNT) light-chain proteases. We evolved BoNT/X protease into separate variants that preferentially cleave vesicle-associated membrane protein 4 (VAMP4) and Ykt6, evolved BoNT/F protease to selectively cleave the non-native substrate VAMP7, and evolved BoNT/E protease to cleave phosphatase and tensin homolog (PTEN) but not any natural BoNT protease substrate in neurons. The evolved proteases display large changes in specificity (218- to >11,000,000-fold) and can retain their ability to form holotoxins that self-deliver into primary neurons. These findings establish a versatile platform for reprogramming proteases to selectively cleave new targets of therapeutic interest.

Pooled CRISPR Screening Informs Rapid Remodeling of the Secretory Pathway in Non-Model Yeast

Dalvie, N. - Presenter, Massachusetts Institute of Technology

Lorgeree, T., Massachusetts Institute of Technology

Love, J., Massachusetts Institute of Technology

The yeast *Komagataella phaffii* (*Pichia pastoris*) is commonly used for the production of recombinant proteins due to its rapid growth to high cell densities and its ability to secrete proteins into the culture medium. Large-scale clinical manufacturers routinely achieve industrially relevant titers for protein therapeutics like insulin and subunit vaccines. There is growing interest in *K. phaffii* as an alternative host for the manufacturing of more complex, high-value protein therapeutics like monoclonal antibodies (mAbs), including several recent product approvals. Early engineering efforts yielded humanized yeast strains capable of secreting mAbs with specific quality attributes and post-translational modifications. Secreted titers must be increased, however, to realize cost-saving or scalability advantages over current hosts like CHO cells.

Yeasts devote a significant portion of cellular resources to protein secretion â€" including several hundred secreted endogenous host-cell proteins (HCPs). Many HCPs are not essential in laboratory or manufacturing conditions, and those resources could be rerouted to production of a recombinant protein. In non-model hosts like *K. phaffii*, however, it is not known which HCPs could be removed without cost to the cells, making large scale genome engineering a tedious effort. Here, we used an unbiased pooled CRISPR knockout library to identify non-essential genes in the yeast secretory pathway. We showed that genes predicted to be non-essential could be rapidly disrupted with little engineering effort. Finally, we evaluated the performance of engineered strains for secretion of several recombinant proteins, including a mAb. This study demonstrates a general strategy for facilitating genomic engineering in non-model organisms using unbiased knockout screens.

Portable Bacterial CRISPR Transcriptional Activation Enables Metabolic Engineering in Multiple Bacterial Species

Kiatissewee, C. - Presenter, University of Washington

Fontana, J., University of Washington

Dong, C., University of Washington

Sugianto, W., University of Washington

Karanjia, A., Arizona State University

Peralta-Yahya, P., Georgia Institute of Technology

Carothers, J., University of Washington

Zalatan, J., University of Washington

CRISPR-Cas transcriptional programming in bacteria is an emerging tool to regulate gene expression for various applications, including metabolic pathway engineering. Here, we developed and optimized new CRISPR-Cas transcriptional activation (CRISPRa) systems in *E. coli* and *P. putida*. We provide a methodology to transfer CRISPRa to a new host by first optimizing expression levels for the CRISPRa system components, and then applying rules for effective CRISPRa based on a systematic characterization of *E. coli* promoters. We observed a correlation between CRISPR-mediated expression levels in *E. coli* and *P. putida* with $R^2 = 0.80$. The CRISPRa system can be implemented simultaneously with CRISPR interference (CRISPRi) for transcriptional repression, which enables complex regulation of multi-gene pathways. We demonstrated that CRISPRa can be used to regulate heterologous and endogenous genes, and can be controlled with small molecule inducible promoters. We used this optimized system to regulate multi-gene biosynthetic pathways for biopterin and mevalonate production, and observed 5-fold and >40-fold increases in product titers with CRISPRa-mediated pathway activation. This work will enable new metabolic engineering strategies in *E. coli* and *P. putida* and pave the way for CRISPR-Cas transcriptional programming in other bacterial species and microbial consortia.

Precise, Single-Cell Optogenetic Induction of Antibiotic Resistance for Selection of Escherichia coli

Blassick, C. M. - Presenter, Boston University

Dunlop, M. J., Boston University

Optogenetic control, or the use of light to control gene expression, is a valuable tool for both elucidating and influencing cell behavior. Advantages of optogenetic tools over chemical induction include high temporal and spatial precision, quickly reversible kinetics, and ease of integration with computational control. Here, we use the CcaSR two-component system, which activates transcription in the presence of green light and represses transcription in the presence of red light, to control expression of a tetracycline resistance gene both in bulk culture and at the single-cell level. We show that cells expressing this system are reliably killed when exposed to red light in the presence of inhibitory levels of tetracycline, while cells exposed to green light continue to grow at a high rate. Using this system in combination with real-time cell segmentation and feature extraction, it is possible to target cells for selection based on desired phenotypic traits, such as size, expression of a fluorophore, or a certain growth rate. We present results demonstrating successful selection based on expression of a fluorophore and progress towards recovering cells from the device, which can then be used for additional rounds of selection or for other purposes. The ability to dynamically target a large number of individual cells based on arbitrary traits will make this system useful for selecting cells based on features that are not easily identified by conventional means, such as dynamic behavior over the course of many hours. Additionally, the high precision and reliability of the described antibiotic resistance induction setup make it well-suited for a number of other potential applications, including exploring the dynamics of antibiotic resistance mechanisms or controlling the growth rate of cultures.

Precision Engineering of Biomolecular Function with Massively Multiplexed Genotype-Phenotype Measurements and Machine Learning

Ross, D. - Presenter, National Institute of Standards and Technology

Tack, D. S., University of Texas at Austin

Tonner, P. D., NIST

Romantseva, E. F., Material Measurements Laboratory, National Institute of Standards and Technology

Alperovich, N., NIST

Vasilyeva, O., NIST

Olson, N. D., NIST

Pressman, A., NIST

Levy, S. F., SLAC National Accelerator Laboratory

For engineering biology to become a mature engineering discipline, there is a need for methods to engineer complex biological functions with quantitative, made-to-specification precision. In this presentation, weâ€™ll describe results for a new approach to biomolecular engineering: Inspired by directed evolution, we start with a large library of biomolecular variants. But, instead of picking a few â€œwinnersâ€ from a laboratory selection, we measure the genotype and corresponding phenotype for every biomolecule in the library across several different chemical environments. This â€œmeasure everythingâ€ approach enables precise biomolecular engineering in two ways: First, we can identify genotypes from the library that satisfy quantitatively targeted phenotypic specifications without the limitations of a laboratory selection. Second, the large datasets, together with interpretable machine learning models, can reveal systematic sequence-structure-function design rules that provide a predictive understanding of the targeted biomolecular function.

In our first demonstration of this approach, we created a library of nearly 100,000 variants of the LacI sensor in *E. coli*. We used laboratory automation and a combination of long- and short-read sequencing to measure the full dose-response curves and corresponding DNA sequences for every variant. With the resulting data, we identified LacI genotypes with precisely targeted dose-response. For example, we engineered sensors with sensitivities (i.e. EC_{50}) spanning 3 orders of magnitude with a 1.25-fold accuracy. In addition, we used the data to train interpretable machine learning models that provide an intuitive route to engineer new LacI variants with quantitatively predictable dose-response.

Remarkably, we also found many LacI variants with phenotypes that differ qualitatively from the wild-type, including inverted dose-response variants and never-before-seen band-stop (on-off-on) variants. These qualitative phenotypic changes are particularly interesting because they can provide specific insight into the biophysics of sensor proteins and because they highlight the capability for large-scale genotype-phenotype measurements to discover rare and useful biomolecule variants.

Programming Multicellular Genetic Circuits and Sequential Logic in Synthetic Microbial Consortia

Zeng, M. - Presenter

Andrews, L. B., University of Massachusetts Amherst

Voke, B.

For biomanufacturing, engineered microbial consortia are a useful strategy to produce biomolecules with high productivities by tailoring each microbial strain to perform a subset of the biocatalytic steps and utilizing dynamic enzyme expression. Moreover, engineered consortia could be utilized in a variety of settings, including for therapeutic purposes, and offer the potential to perform complex metabolic and regulatory functions that exceed those of an individual strain. Dynamic regulation and programmable responses to specified cues within consortia are of great interest to implement these functions robustly. To achieve dynamic control for a user-defined application, designable genetic circuits can be employed. While robust and scalable algorithms to design genetic circuits in single cells have been developed, similar algorithms to design consortia-level genetic circuits has lagged behind.

In this work, we present a platform for the design of multicellular genetic circuits. We developed a new set of characterized genetic parts for intercellular communication using quorum sensing components. We tested the incorporation of these cell-cell communication modules and extending the signal matching algorithm for the predictive design of multicellular genetic circuits within engineered consortia. A library of cell-cell communication modules was developed that includes 5 quorum sensors, 5 corresponding signal synthases, and 3 quorum quenching enzymes for autonomous and reversible communication within consortia. Through further engineering, the crosstalk within the sensor library was largely reduced. Quorum quenching enzyme activity and substrate specificity were characterized for the set of cell-cell signals. To validate the design algorithm for multicellular genetic circuits, we constructed consortia comprising two- or three-strains that contain partitioned sequential logic circuits that contain memory and would be useful in diagnostic contexts. Complex combinational logic circuits that could not be implemented in single cell due to the lack of genetic components and metabolic burden were also constructed in consortia.

Protocell Arrays for Simultaneous Detection of Diverse Analytes

Zhang, Y. - Presenter, Georgia Institute of Technology

Kojima, T., Georgia Institute of Technology and Emory School of Medicine

Kim, G. A., Georgia Institute of Technology

McNerney, M., Georgia Institute of Technology

Takayama, S., Georgia Institute of Technology

Styczynski, M. P., Georgia Institute of Technology

Simultaneous detection of multiple analytes from a single sample (multiplexing), particularly when at the point of need, can guide complex decision-making without increasing the required sample volume or cost per test. Despite recent advances, multiplexing still typically faces the critical limitation of measuring only one type of molecule per assay platform â€" for example, only small molecules or only nucleic acids. In this work, we address this bottleneck with a customizable platform that integrates cell-free expression (CFE) with a polymer-based aqueous two-phase system (ATPS) to produce membrane-less â€œprotocellsâ€� containing transcription and translation machinery used for analyte detection. Multiple protocells are arrayed in microwells where each protocell droplet performs distinct reactions to detect chemically diverse targets, including small molecules, minerals, and nucleic acid sequences, all from the same sample. We demonstrate that these protocell arrays can measure analytes in a human biofluid matrix, maintain function after lyophilization and rehydration, and produce visually interpretable readouts, illustrating its potential for application as a minimal-equipment, field-deployable, multi-analyte detection tool.

Rapid Characterization of Spike Variants Via Mammalian Cell Surface Display

Javanmardi, K. - Presenter, University of Texas at Austin

The SARS-CoV-2 spike (S) protein is a critical component of subunit vaccines and a target for neutralizing antibodies. Spike is also undergoing immunogenic selection with clinical variants that increase infectivity and partially escape convalescent plasma. Here, we describe spike display, a high-throughput platform to rapidly characterize glycosylated spike ectodomains across multiple coronavirus-family proteins. We assayed ~200 variant SARS-CoV-2 spikes for their expression, ACE2 binding, and recognition by thirteen neutralizing antibodies (nAbs). An alanine scan of all five N-terminal domain (NTD) loops highlights a public class of epitopes in the N1, N3, and N5 loops that are recognized by most of the NTD-binding nAbs. Some clinical NTD substitutions abrogate binding to these epitopes but are circulating at low frequencies around the globe. NTD mutations in variants of concern B.1.1.7 (United Kingdom), B.1.351 (South Africa), B.1.1.248 (Brazil), and B.1.427/B.1.429 (California) impact spike expression and escape most NTD-targeting nAbs. However, two classes of NTD nAbs still bind B.1.1.7 spikes and neutralize in pseudoviral assays. B.1.351 and B.1.1.248 include compensatory mutations that either increase spike expression or increase ACE2 binding affinity. Finally, B.1.351 and B.1.1.248 completely escape a potent ACE2 peptide mimic. We anticipate that spike display will accelerate antigen design, deep scanning mutagenesis, and antibody epitope mapping for SARS-CoV-2 and other emerging viral threats.

Rapid Generation of Potent Antibodies By Autonomous Hypermutation in Yeast

Authors:

Wellner, A. - Presenter, UC Irvine

The predominant approach for antibody generation remains animal immunization, which can yield exceptionally selective and potent antibody clones owing to the powerful evolutionary process of somatic hypermutation. However, animal immunization is inherently slow, has poor compatibility with certain antigens (*e.g.*, integral membrane proteins), and suffers from self-tolerance and immunodominance, which limit the functional spectrum of antibodies that can be obtained. Here, we describe Autonomous Hypermutation yEast surfAce Display (AHEAD), a synthetic recombinant antibody generation technology that imitates somatic hypermutation inside engineered yeast. In AHEAD, antibody fragments are encoded on an error-prone orthogonal DNA replication system, resulting in *Saccharomyces cerevisiae* populations that continuously mutate surface-displayed antibody repertoires. Simple cycles of yeast culturing and enrichment for antigen binding drive the evolution of high-affinity antibody clones in a readily parallelizable process that takes as little as 2 weeks. We applied AHEAD to generate nanobodies against the SARS-CoV-2 S glycoprotein, a GPCR, and other targets. The SARS-CoV-2 nanobodies, concurrently evolved from an open-source naÃ¯ve nanobody library in 8 independent experiments, reached subnanomolar affinities through the sequential fixation of multiple mutations over 3-8 AHEAD cycles that saw â^¼580-fold and â^¼925-fold improvements in binding affinities and pseudovirus neutralization potencies, respectively. These experiments highlight the defining speed, parallelizability, and effectiveness of AHEAD and provide a template for streamlined antibody generation at large with salient utility in rapid response to current and future viral outbreaks.

Rapid Synthetic Viral Vectorology for Future Pandemic Response

Hewitt, F. C. - Presenter, Signature Science, LLC

Kappell, A. D., Signature Science, LLC

Scholes, A. N., Signature Science, LLC

Keplinger, N. C., Signature Science, LLC

Characterizing viral biothreats and predicting potential evolutionary pathways is crucial to mitigate future pandemics. Approaches utilizing *in silico* modeling are ultimately limited by the lack of empirical data and the immense scale of the theoretical evolutionary landscape. Conversely, laboratory experiments which aim to study viral biothreats are effective but must be performed at high biosafety levels or chimeric viruses. To overcome these challenges, we have employed a multidisciplinary approach combining viral vectorology and basic virological study to establish RAVEN (Rapid Attenuated Virus EvolutioN). In the same way that viral vector gene therapy approaches replace viral genes with a genetic â€œpayload,â€� RAVEN similarly replaces critical replication genes with suitable reporter genes. The excised viral genes are then expressed *in trans,* permitting viral vector assembly. The resulting viral particles are infectious and essentially identical to wild type but cannot replicate. For this reason, viruses requiring high biocontainment can be propagated in a Biosafety Level-2 environment. While the aim of RAVEN is to establish a broad toolkit applicable to a wide variety of viruses, our current work has focused on the establishment of a vector model of SARS-CoV-2. We have produced a synthetic SARS-CoV-2 vector which lacks key replication genes. These open reading frames have been replaced with reporter genes which were stitched into the polyprotein sequence. All homology between the *transgenes* and vector was removed to prevent recombination that could restore a replication-competent virus. Resulting vector particles are infectious and are capable of GFP transduction of cells. Critically, these vectors are not replication competent in the absence of *transgene* expression and the absence of transgene RNA in viral vector preparations has been confirmed. We believe this approach supports safer, faster characterization of existing and future viral threats and will serve as a critical tool to reduce the impacts of future viral pandemics.

Rational Engineering of a Modular Bacterial CRISPR-Cas Activation Platform with Expanded Target Range

Villegas Kcam, M. C. - Presenter, Rice University

Chappell, J., Rice University

Tsong, A. J., Rice University

The ability to control gene expression is a transformative capability that allows to uncover gene function or induce valuable phenotypes and cell fates important for applications in biotechnology and biomedicine. CRISPR-Cas regulators have provided an excellent toolbox for this purpose. Among these, CRISPR activator (CRISPRa) systems turn on transcription through localization of activation domains (AD). However, current bacterial CRISPRa systems still present several challenges. First, different bacterial ADs offer distinct regulatory properties, meaning different AD are often required depending on the target gene or desired application. Second, the bacterial CRISPRa systems need to be targeted to sites located a specific number of nucleotides upstream of the promoter to achieve activation, which critically limits its application in non-synthetic targets. To address these limitations, we adopt a protein engineering approach to create a highly modular and target flexible bacterial CRISPRa system. We identified a novel and strong AD which we harness to explore AD recruitment strategies. Specifically, we evaluated the use of modular protein-protein interaction domains which allows different AD to be encoded on independent and easily exchangeable plasmid elements. Additionally, we evaluated the use of circularly permuted variants of the dCas9 protein to identify variants capable of activating gene expression from distinct positions previously identified as non-activating sites and therefore expand the functional target range. Our modular CRISPRa system provides a more flexible and versatile tool that will expand our ability to activate and control bacterial genomes.

Reaping Sweet Benefits of Industrialized Synthetic Biology: A Case Study of Successful Manufacturing Scale-up for Rebaudioside M

Khankhoje, A. - Presenter

Wichmann, G., Amyris

Synthetic biology is a rapidly developing industry interfacing Chemistry, Biology and Computer Science. The goal is to make efficacious, pure products faster, cheaper, and more sustainably. Because the products are made by yeast, a living organism, many more new molecules can now be made without traditional chemical synthesis. Yet, there are different limitations and new challenges to manufacturing. Amyris is a science and technology leader in the research, development and production of sustainable ingredients which applies its innovative bioscience platforms to convert plant sugars into a wide variety of molecules. Amyris is also one of the very few synthetic biology companies that has successfully managed to bridge the â€œvalley of deathâ€� from lab-to-market, commercializing 13 molecules.

The talk will focus on a new product Amyris has developed, namely the zero-calorie sweetener Rebaudioside M (RebM). RebM is found naturally at low concentrations in the plant *Stevia rebaudiana*, from which the alternative sweetener stevia is derived. Depending on the category of application, RebM is approximately 300 times sweeter than sucrose and does not have the metallic or licorice flavors typically associated with stevia extracts. In fact, the taste profile of RebM is so remarkably similar to sucrose that RebM is the most desirable zero-calorie sweetener on the market today. However, RebM comprises < 1% of the total steviol glycosides produced by the stevia plant, making purification prohibitively difficult and expensive. Amyris has engineered a yeast strain to convert cane sugar to pure RebM, and in December of 2018 Amyris launched RebM as a new commercial ingredient. In December 2019, Amyris launched its own consumer brand, Purecane with RebM as its hero ingredient. Since its launch, the Purecane brand has earned a number of awards, including *Popular Science*, Top 100 Greatest Innovations of 2020 and *Fast Company,* Most Innovative Brand of 2021.

Recombinant Cell-Free Extracts for Biocatalytic Cascades: Fertilizers to Diagnostics

Alissandratos, A. - Presenter, Australian National University

The ongoing COVID-19 pandemic has highlighted the need for timely, cost-effective, and accurate diagnostics. It has also revealed how shortages of plastic consumables, buffers, and PPE can limit traditional assays. Sample collection and centralized testing with RT-qPCR has dramatically improved the time to result, but it remains too slow to mitigate the spread of infection. To help address these issues, we have developed a frugal SARS-CoV-2 diagnostic enabled by cell-free synthetic biology circuitry coupled to a paper-based resonant sensor that can be safely and discretely scanned in a closed envelope and has the potential for at-home use and distributed contact-free testing. Lysed samples undergo isothermal reverse transcription recombinase polymerase amplification (RT-RPA) in a sample well which is placed onto a lyophilized, paper-based cell-free expression system. The cell-free system contains plasmids that harbor toehold switches selective to specific targets on the amplified RPA products. Complementary viral RNA amplicons complex with the toehold switches to activate translation of subtilisin, a strong protease reporter. The protease then degrades gelatin switches coated on the paper, which overlay a screen-printed paper-based resonant (LC) sensor. The sensor has a resonant frequency which shifts due to changes in permittivity caused by gelatin degradation. This clear frequency change can be measured rapidly through a closed envelope using an embedded portable vector network analyzer. We envision the technology can be deployed with autonomous scanners positioned at local drive-up lanes, hospitals, or workplaces and would be capable of signaling within 10 seconds if the closed sample envelope contains a positive or negative sample. This workflow of individualized administration followed by rapid, local scanning would eliminate the current problems of test site back-log, lack of PPE, and reagent scarcity and would be well-suited for deployment against a variety of existing and emerging pathogens for outbreak tracking and pandemic preparedness.

Redesign of Lactococcus Lactis genome for Increasing pDNA and Recombinant Protein Production Using a CRISPR/Cas9 Strategy

Duarte, S. - Presenter, Institute for Bioengineering and Biosciences, Instituto Superior Técnico

Monteiro, G., Institute for Bioengineering and Biosciences, Instituto Superior Técnico

The degradation of recombinant plasmid DNA by endonucleases is one of the major bottlenecks that is keeping *Lactococcus lactis* from reaching its full potential as cell factories for pDNA and recombinant protein production with pharmaceutical applications.

Escherichia coli has been the gold standard, since it is able to achieve high pDNA and protein yields, but *L. lactis* is a promising alternative, with its food-grade status. The purification process itself is already more cost effective for *L.* lactis, due to the absence of co-purifying lipopolysaccharides, a good secretion system and the presence of only one major extracellular protease. If endonuclease genes could be removed, the pDNA and, consequently, protein yields, should increase and turn *L. lactis* in a laboratory and industrial work horse.

There are not easy and effective systems for editing genes in *L. lactis*, so we decided for the CRISPR/Cas9 strategy, for a scarless and targeted removal. The present work consists in the adaptation of the *Lactobacillus casei* pLCNICK one-step system to *L. lactis*.

The *nth* gene was chosen as the first knockout target, since it codes for an endonuclease III that could be responsible for creating gaps in the plasmid vectors, contributing for its quick degradation. The pLCNICK is currently being engineered in order to Cas9 and the homologous recombination system be able to recognize and knockout the *nth* gene.

Several *L. lactis* strains were transformed with the designed plasmid and the *nth* knockout confirmed by PCR amplification and sequencing. After plasmid curing, the strains were transformed with a GFP reporter plasmid and the effect of the *nth* gene knockout in the plasmid yield and quality analyzed by agarose gel electrophoresis, spectrophotometry and chromatography of the purified pDNA, fluorimetry, fluorescence microscopy and flow cytometry using whole cells. Real-time quantitative PCR was used to quantify the plasmid copy number.

Redesign of Metabolic Pathways for High Yield Production of ?-Carotene in Yarrowia Lipolytica

Pappalardo, X. - Presenter, University of Catania

Jansen, G., University of Cambridge

Amaradio, M. N., University of Catania

Costanza, J., Italian Institute of Technology

Nicosia, G.

Biotechnological synthesis of Î²-carotene for its nutraceutical and antioxidant properties is becoming more attractive and aims to enhance the production using novel recombinant Î²-carotene *producing yeast* strains. The use of the engineered yeast *Yarrowia lipolytica* as a competitive producer organism of Î²-carotene, is the current method to naturally sustain the high supply of Î²-carotene in large-scale fermentative production. Towards this goal, we performed several simulations for the identification of competitive *in silico* strains of *Y. lipolytica* because of selected simulated in gene deletions within its metabolic network, which was adapted to mimic an engineered yeast capable of producing Î²-carotene. The design BioCad algorithm is then able to explore the solution space of the problem, defined by all the possible combinations of gene deletions, optimizing multiple objectives such as growth rate, target production or productivity as predicted by the *flux balance analysis*. The framework, based on an ad-hoc optimization algorithm for the consequential selection of the notable gene deletions, identifies the most promising sub-optimal strains within the Pareto Fronts. The strains selected by the procedure undergo further characterization, such as a selection of strains with the more appropriate genes to be deleted at the same time, with a more efficient usage of nutrients or with less waste metabolites produced. Our pipeline for in silico identification of competitive strains of engineered *Y. lipolytica* through the selection of effective gene deletions that can test and rank a high number of candidates. This approach is effective for the optimization of the design of competitive strains and is also helpful for accelerate the experimental processes of predicted mutants in vitro. Î²-carotene production was obtained on a wide range of different media and with different combinations of carbon sources.

Screening for Microbial Therapeutic Interactions in Malignancy

Ricaurte, D. - Presenter, Columbia University

Huang, Y., Columbia University

Wang, H., Columbia University

Bacteria comprising the human gut microbiome have been linked to a myriad of health outcomes, including response to cancer treatment. Although an association between the microbiome and chemotherapeutic agents is well established, mechanisms underlying â€œbug-drugâ€� interactions remain poorly understood. In large part, this is due to bacterial strain-level variation, as well as the difficulty of identifying and isolating strains that interact with chemotherapeutics from the larger gastrointestinal bacterial community. We hypothesized that improved bacterial growth in the presence of cancer therapeutics is a valid marker of a significant bacterial-drug interaction. To this end, we have developed a novel microbe-centric screening method to identify and isolate bacterial strains in human microbiomes that interact with cancer agents of interest, with the ultimate goal of identifying the bacterial genetic elements responsible for the observed growth shifts. In preliminary studies, over forty therapeutics have been screened in three different human microbiota. Our results suggest that select agents have a reproducible and significant growth impact, driven by previously unknown antimicrobial effects, on a subset of microbial strains. We have successfully isolated sensitive and resistant strains of interest from human microbiota; comparative genomic, transcriptomic, and directed evolution studies to identify relevant genetic elements are ongoing. If successful, this work could result in the identification of novel pathways in the antibiotic resistome that carry meaningful implications in the context of cancer treatment, and which could be leveraged more broadly in future efforts to engineer the gut microbiome. More broadly, our microbiome screening pipeline may be applied to a range of drug classes to allow for discovery of not only genetic response mechanisms, but also identification of regulators of gene expression. These xenobiotic compounds could be repurposed as chemical modulators for programmable induction or repression of gene expression of gut bacteria in future bioengineering and synthetic biology applications.

Single Cell Chemical Imaging with Stimulated Raman Scattering for Biofuel Production Screening

Tague, N. - Presenter, Boston University

Lin, H., Boston University

Shin, J., MIT

Zhang, J., Boston University

Lugagne, J. B., Boston University

Wong, W., Boston University

Cheng, J. X., Boston University

Dunlop, M. J., Boston University

The convergence of metabolic engineering and synthetic biology disciplines has led to increasingly efficient microbial production of valuable chemicals such as biofuels. Recent advances in genetic and protein engineering allow for more extensive strain optimization and diverse chemical production. However, state of the art screening techniques are destructive and lose single cell production data. Stimulated Raman scattering (SRS) imaging has the potential to be used as a single cell screening platform by directly detecting chemicalsâ€™ Raman spectra. Here, we demonstrate the ability to differentiate production levels of several biofuels within *E. coli* based on unique Raman spectra footprints. Utilizing the spectral information, we showcase the ability to differentiate free fatty acids of different chain lengths, allowing for accurate enzyme specificity prediction. Lastly, we demonstrate the ability to perform longitudinal SRS imaging to study production dynamics. The results show the potential of this imaging platform to be used to guide strain optimization for metabolic engineering projects.

Single-Cell and Single-Transcript Metrology for Engineering Biology

Rammohan, J. - Presenter, NIST

Lund, S., NIST

Alperovich, N., NIST

Paralanov, V., NIST

Strychalski, E. A., NIST

Ross, D., National Institute of Standards and Technology

Single-cell and single-transcript measurement methods have elevated our ability to understand and engineer biological systems. However, defining and comparing performance between methods remains a challenge, in part due to the confounding effects of experimental variability. Here, we propose a generalizable framework for performing multiple methods in parallel using split samples, so that experimental variability is shared between methods. We demonstrate the utility of this framework by performing 12 different methods in parallel to measure the same underlying reference system for engineered cellular response. We compare method performance using quantitative evaluations of bias and resolvability. We attribute differences in method performance to steps along the measurement process such as sample preparation, signal detection, and choice of measurand. Finally, we demonstrate how this framework can be used to benchmark a new method for single-transcript detection. The framework we present here provides a practical way to compare performance of any methods.

Small-Molecule Biosensors for High-Throughput Screening and Control of Cellular Metabolism

Authors:

Jensen, M. K. - Presenter, Technical University of Denmark

Petersen, S., Novo Nordisk Foundation Center for Biosustainability

Zhang, J., Technical University of Denmark

D'Ambrosio, V., Technical University of Denmark

Lengger, B., DTU Biosustain

Damgaard Jensen, E., Technical University of Denmark

Elise Hoch-Schneider, E., Technical University of Denmark

Engineering living cells to perform new-to-nature functions require both understanding of multiple interacting genes and cellular processes, and levers by which they can be independently screened and controlled for optimal performance. Allosterically regulated transcription factors and membrane-bound receptors have proven widely applicable in synthetic biology as ligand-specific biosensors enabling real-time monitoring, selection and dynamic regulation of cellular metabolism.

This presentation will initially focus on methods for rational and evolution-guided engineering of biosensors with user-defined transfer functions and small-molecule specificities. Next, applications of biosensor for screening, selection and control of cellular metabolism in the eukaryote model chassis, bakerâ€™s yeast *Saccharomyces cerevisiae*, will be presented. This includes examples on high-throughput screening of biosynthetic pathways, selection based on conditional growth, and robust safe-guarding of refactored metabolic pathways from evolutionary drifting. Lastly, the power of using real-time biosensor read-outs from diverse cellular designs to train machine learning algorithms for predictive engineering of cellular metabolism will be presented.

Taken together, the results presented leverage the modular design protein-based biosensors to enable monitoring, control, and predictive engineering of cellular metabolism.

Strategies for Quantifying and Enhancing Genetic Code Manipulation in Yeast

Authors:

Van Deventer, J. - Presenter, Tufts University

Stieglitz, J. T., Tufts University

Potts, K., Tufts University

Zackin, M., Tufts University

Lei, M., Tufts University

Noncanonical amino acids (ncAAs) facilitate a diverse range of applications in protein science and engineering ranging from dissecting key protein functions to engineering proteins with more â€œdruglikeâ€ properties. Genetically encoding ncAAs in proteins produced within cells and in vitro provides the basis for preparing ncAA-containing proteins that would be difficult or impossible to produce via chemical synthesis. However, our understanding of the performance of these systems remains limited, especially from a quantitative perspective. Here, we present our efforts to (1) establish tools to measure ncAA incorporation in *Saccharomyces cerevisiae* in response to the amber stop codon; and (2) use these tools to enhance ncAA incorporation efficiency and fidelity in yeast. We have identified a series of dual reporter systems that support flow cytometry-based measurements of ncAA incorporation efficiency and fidelity. These reporters enable quantitative measurements for a diverse range of orthogonal translation systems (OTSs) and for a wide range of yeast strains. We surveyed a number of previously described OTSs in yeast and determined that these systems typically support ncAA incorporation at approximately 15-20% of wild-type protein translation events. Many of these OTSs also exhibit polyspecificity (incorporation of several ncAAs). Ongoing work in the laboratory seeks to use these reporter systems to identify OTSs with improved or altered ncAA incorporation. We have also initiated studies in which we use our reporters to screen a broad range of yeast strains for phenotypes of enhanced ncAA incorporation. Our early findings demonstrate the feasibility of this approach using single-gene knockout strains with known alterations in stop codon readthrough. The measurement tools we have developed form the basis for evolving yeast to better accommodate alternative genetic codes.

Synthesis and Characterization of Biogenic Selenium Nanoparticles Made from Pathogenic Bacteria with Selective Antimicrobial Properties

Truong, L. - Presenter, Northeastern University

Medina, D., Northeastern University

Antimicrobial Resistance has become an urgent crisis for humanity. An alternative, non-resistant-inducing and effective way to treat infection is needed. Metallic Nanoparticles (NPs) have long been studied for their antimicrobial properties, yet cost, efficiency, processing difficulties and ecological safety prevent their clinical use. To remedy these challenges, biogenic synthesis of nanoparticles, specifically Selenium nanoparticles (SeNPs) are explored and presented as an alternative antimicrobial agent. Selenium NPs was synthesized using E. coli, P. aeruginosa, S. aureus and MRSA with final diameter of 80-120 nm by culturing bacteria cells in standard conditions and inoculating them with Selenium salt at the end of exponential growth phase. The nanostructures were characterized using Transmission Electron Microscopy (TEM) and Energy Dispersive X-Ray (EDX) to determine the chemical compositions and Inductively coupled plasma mass spectrometry (ICP-MS) to validate the chemistry within the samples. Nanoparticles were tested against E. coli, P. aeruginosa, S. aureus and MRSA for their ability to inhibit the bacterial growth through optical density measurements and colony forming unit assays. Biocompatibility tests of the nanoparticles with human HDF were accomplished, along with resistance studies by several rounds of exposure onto pathogens. Results indicate that SeNPs were successfully generated by all tested strains. The biogenic SeNPs exhibited antimicrobial properties, with log reduction ranging from 3 â€" 4 with crossed application and 5 â€" 6 with specific application. Cytotoxicity assays confirmed its safety profile against HDF cells. More significantly, biogenic SeNPs treatment did not induce resistance, in contrast with commercial penicillin and commercialized AgNPs. These findings confirm the viability of biogenic SeNPs as an easy, safe, cheap and effective treatment and a potential solution against the Antimicrobial Resistance Crisis.

Synthetic Biology Approaches for the Optimization and Improvement of a Live Bacterial Therapeutic for the Treatment of Phenylketonuria (PKU)

Monahan, C. - Presenter, Synlogic

Phenylketonuria (PKU) is a human metabolic disease characterized by an inability to degrade phenylalanine (Phe), causing neurotoxicity. A novel therapeutic treatment to consume Phe in the GI tract, *Escherichia coli* Nissle (EcN) SYNB1618 was engineered to express two independent enzymes, PAL and LAAD, which metabolize Phe into the respective metabolites, *trans*-cinnamic acid (TCA) and phenylpyruvate (PP). Uptake of Phe is facilitated by a high efficiency transporter, PheP. In a Phase 1 trial, SYNB1618 led to dose-dependent increases in SYNB1618-specific biomarkers in healthy humans.

To further develop our Synthetic Bioticâ„¢ platform, a more potent PKU strain was sought through optimization of whole cell PAL activity. *In vitro* analyses suggested the rate-limiting step was feedback inhibition of PAL by TCA. Increasing PAL expression did not increase whole cell activity but did increase activity in lysate, while salicylate, an efflux pump inducer, increased whole-cell PAL activity.

Using Zymergenâ€™s proprietary sensor engineering platform, an allosteric transcription factor biosensor controlling GFP expression was developed which responded specifically to TCA in a dose-dependent manner. Biosensor intensity correlated with PAL activity as measured by GFP and TCA production. Utilizing a directed evolution approach, PAL variants were designed for improved activity using phylogenetic analysis combined with computational protein design algorithms. Iterative screening of oligo-based combinatorial libraries resulted in progressive activity improvement over multiple rounds of enrichment. Unique PAL variants with increased activity were identified and characterized. The best-performing variants displayed increases in K_m and V_{max}.

SYNB1934 was constructed by integrating four copies of a lead PAL variant into the chromosome of EcN modified with key features of SYNB1618. *In vitro*, SYNB1934 exhibited a 2-fold increase in TCA production rate compared to SYNB1618. In healthy non-human primates, SYNB1934 treatment resulted in increases of strain activity biomarkers *in vivo*. These results define a strategy for optimizing live bacterial therapeutics.

Synthetic Epigenetic Reader Engineering with an All-in-One Cell-Free Expression and Protein Interaction Microarray

Authors:

Harrison, J. - Presenter, Emory University

Rafat, N., Michigan State University

Sarkar, A., Georgia Tech

Haynes, K., Emory University

Chromatin, the system of genomic DNA, RNA, and proteins that organize chromosomes, is misregulated in cancer and has therefore become the target of an emerging class of epigenetic cancer therapies. These treatments rely heavily on small compounds to interfere with chromatin protein activity, while the vast potential of protein engineering to modulate chromatin in cancer has barely been explored. Our vision is to engineer synthetic reader-actuators (SRAs), customizable chromatin proteins that â€œreadâ€� chromatin features and modulate tumor suppressor and oncogene expression. Our immediate goal is to design an SRA with enhanced binding affinity for a cancer associated histone post-transcriptional modification, H3K27me3, using variants of the polycomb chromodomain (PCD) from the protein CBX8. So far, we achieved a 2-fold increase in SRA affinity for H3K27me3 by using two tandem wildtype PCDs (S. Tekel 2017 ACS Syn. Biol.). To enhance the affinity of a single PCD module, we will screen a library of variants that carry mutations at five amino acid positions that coordinate binding with the histone H3 tail. We developed a Golden Gate cloning platform (J.H. Priode 2021 protocols.io) to rapidly build a library of SRA variants in a cell-free transcription-translation vector (GGDestX1-Amp). We observed 50% Golden Gate insertion efficiency in GGDestX1-Amp, with zero misassembled products, and cell-free expression from GGDestX1-Amp in liquid phase reactions. We have designed a platform called the chromatin nucleic acid-programmable protein array (ChNAPPA), to efficiently screen all 3072 SRA constructs. This platform prints SRA-encoding DNA onto poly-L-lysine coated glass slides, alongside H3K27me3 histone tail peptides. After cell-free expression, the high-affinity variants are captured by the histone tail â€œbaitsâ€� within each microspot. High-affinity SRA candidates from ChNAPPA will be tested for gene-activating function, compared to our standard SRAs (1x and 2x wildtype PCD-VP64), at silenced tumor suppressor genes in cancer cells *in vitro*.

Synthetic Promoter Design in Escherichia coli Based on a Deep Generative Network

Wang, Y. - Presenter, Tsinghua University | Dept of Automation

Wang, H., Tsinghua university

Wei, L., Tsinghua university

Liu, L., Tsinghua university

Wang, X., Building of biotechnology

Promoter design remains one of the most important considerations in metabolic engineering and synthetic biology applications. Theoretically, there are 4^{50} possible sequences for a 50-nt promoter, of which naturally occurring promoters make up only a small subset. To explore the vast number of potential sequences, we report a novel AI-based frame- work for de novo promoter design in *Escherichia coli*. The model, which was guided by sequence features learned from natural promoters, could capture interactions between nucleotides at different positions and design novel synthetic promoters *in silico*. We combined a deep generative model that guides the search for artificial sequences with a predictive model to preselect the most promising promoters. The AI-designed promoters were optimized based on the promoter activity in *E. coli* and the predictive model. After two rounds of optimization, up to 70.8% of the AI-designed promoters were experimentally demonstrated to be functional, and few of them shared significant sequence similarity with the *E. coli* genome. Our work provided an end-to-end approach to the de novo design of novel promoter elements, indicating the potential to apply deep learning methods to de novo genetic element design.

Systematically Delivering Biodegradable Mesoporous Silicananoparticles Via Dissolvable Microneedles to Promote Remyelination in Patients with Neurodegenerative Disease

Kalra, A. - Presenter, Carmel High School

Sriram, R. - Presenter, Carmel High School

Chitturi, G. - Presenter, Carmel High School

Jayarapu, K. - Presenter, Carmel High School

Nanoparticle delivery via microneedle patch systems is an upcoming area of research with promising results and broad applications. Using an epidermal patch with microneedles to deliver nanoparticles and small molecules and target them to different locations in the body can allow for micromanipulation and nanoscale treatment plans that involve multiple interventive strategies, especially in patients with neurodegenerative diseases. Here we discuss potential ways in which polymeric nanoparticles can enter the peripheral nervous system as a method to treat different neurodegenerative diseases. We specifically highlight Guillain-BarrÃ© Syndrome as a case study of our method and show how nanoparticles can be used to upregulate vital proteins such as NGR1 and Adam22 to neurons and modify their gene expression to increase Schwann cell myelination activity. This novel technique can help the lives of countless people affected by neurodegenerative diseases worldwide.

Temperature-Responsive Optogenetic Probes of Cell Signaling

Bugaj, L. - Presenter, University of Pennsylvania

Benman, W., University of Pennsylvania - Philadelphia, PA

Berlew, E., University of Pennsylvania

Deng, H.

Parker, C., University of Pennsylvania

Kuznetsov, I., University of Pennsylvania

Siekmann, A., University of Pennsylvania

Chow, B., University of Pennsylvania

Lim, B., University of Pennsylvania

Optogenetics provides a powerful toolset to dissect the role of signal dynamics in cells and organisms. However, such tools can be challenging to implement because they often consist of multiple components that require stoichiometric tuning and occupy valuable fluorescence channels. We thus sought to create single-component variants of optogenetic probes that work through membrane recruitment, a common mode of optogenetic control. Specifically, we engineered control of Ras/Erk and PI3K/Akt signaling by fusing pathway activators to BcLOV4, a recently characterized fungal photoreceptor that translocates to membrane lipids upon photoactivation. Our probes allowed robust activation of both pathways and compared favorably to existing 2-component probes in terms of dynamic range of induction and minimal perturbation of cell physiology. However, we discovered an unanticipated temperature sensitivity of the BcLOV4 protein, such that membrane localization and pathway stimulation spontaneously decay within ~1 hour â€" despite constant illumination â€" at a rate proportional to both the temperature and the intensity of light. We systematically characterized this unique dependency on both light and temperature, and we developed a computational model that fully predicts BcLOV4 translocation and Ras/Erk activation dynamics across a range of physiological temperatures and light conditions. In addition to use in mammalian cells, we show that BcLOV4-based tools function robustly and stably in both Drosophila and zebrafish, which are not amenable to stoichiometric expression tuning, and which are cultured at temperatures below 30 Â°C where BcLOV4 can be stably stimulated. Our work provides a new set of easy-to-use optogenetic probes that work across organisms, and a quantitative framework for their optimal use. Moreover, BcLOV4 provides a natural substrate for engineering a new class of tools that respond to light and/or temperature.

the Promoter Calculator - a Sequence-to-Function Biophysical Model of Transcriptional Initiation for Sigma70 Promoters with Any Sequence

La Fleur, T. - Presenter, Penn State

Salis, H. M., Pennsylvania State University

Hossain, A., Penn State University

Engineering synthetic promoters with precision control has remained a challenge due to our inability to predict how a promoter's sequence and DNA context controls its function and mRNA output. Here, we developed an accurate sequence-to-function model of transcriptional initiation enabling the automated design of synthetic promoters and the *a priori* prediction of cryptic promoters - both of which are needed to advance Synthetic Biology towards genome-scale functional design. To do this, we combined oligopool synthesis, library-based cloning, and next-generation sequencing to construct and characterize 14,206 rationally designed sigma^{70} promoters *in vitro*. Measurements include transcriptional start site frequencies and overall mRNA levels. This approach enables highly-parallel characterization of constitutive promoter activity without confounding factors such as unintentional transcriptional regulation, non-sigma^{70} activity, and mRNA decay. These measurements, in combination with machine learning, were used to parameterize a thermodynamic model quantifying the interactions controlling transcription rate. For demonstration, the â€œPromoter Calculatorâ€� was used to accomplish three major tasks â€" accurate prediction of thousands of sigma^{70} promoters across various conditions, the *de novo* design of novel sigma^{70} promoters, and the identification of cryptic promoters internal to a genetic circuit. 4,350 highly non-repetitive promoters and 6,165 genome-integrated promoters characterized *in vivo* were accurately predicted by the model with Spearman Rank-Order Coefficients of .68 and .78, respectively. Promoters designed *de novo* using the Promoter Calculator exhibited a 683-fold range in expression *in vivo* resulting in a Pearson Correlation Coefficient of .85. The Promoter Calculator was used to analyze a genetic circuit containing 11 circuit promoters and 29 cryptic promoters. The model identified 29 sigma^{70} promoters out of 40 observed promoters (72.5% accurate), including 10/11 circuit promoters and 19/29 cryptic promoters. The Promoter Calculator facilitates context-aware, rational promoter design without relying on a fixed table of pre-characterized sequences while serving as a powerful tool in promoter identification.

The Tryptophan-Induced tnaC Ribosome Stalling Sequence Exposes High Amino Acid Cross-Talk That Can be Mitigated By Removal of NusB for Higher Orthogonality

Sherman, M. - Presenter, University of Texas at Austin

Contreras, L. M., The University of Texas at Austin

Sandeep, S., University of Texas at Austin

A growing number of engineered synthetic circuits have employed biological parts coupling transcription and translation in bacterial systems to control downstream gene expression. One example, the leader sequence of the tryptophanase (*tna*) operon, is a commonly employed L-tryptophan inducible circuit controlled by ribosome stalling. While induction of the *tna* operon has been well-characterized in response to L-tryptophan, cross-talk of this modular component with other metabolites in the cell, such as other naturally-occurring amino acids, have been less explored. In this study, we investigated the impact of natural metabolites and *E. coli* host factors on induction of the *tna* leader sequence. To do so, we constructed and biochemically validated an experimental assay using the *tna* operon leader sequence to assess differential regulation of transcription elongation and translation in response to L-tryptophan. Operon induction was then assessed following addition of each of the 20 naturally-occurring amino acids to discover that several additional amino acids (e.g. L-alanine, L-cysteine, L-glycine, L-methionine, and L-threonine) also induce expression of the *tna* leader sequence. Following characterization of dose-dependent induction by L-cysteine relative to L-tryptophan, the effect on induction by single gene knockouts of protein factors associated with transcription and/or translation were interrogated. Our results implicate the endogenous cellular protein, NusB, as an important factor associated with induction of the operon by the alternative amino acids. As such, removal of the *nusB* gene from strains intended for tryptophan-sensing utilizing the *tna* leader region reduces amino acid cross-talk, resulting in enhanced orthogonal control of this commonly used synthetic system.

Time-Course Expression Profiling Reveals Early and Late Responders to a Synthetic Chromatin Regulator

Enwerem-Lackland, I. - Presenter, Emory University

Williams, N. L., Emory University

Plaisier, C., Arizona State University

Haynes, K., Emory University

Epigenetic therapy, to activate non-mutated tumor suppressors in cancer cells, is gaining traction as an alternative to traditional chemotherapies that inhibit DNA replication and repair. Currently, epigenetic treatments for cancer are limited to small molecules that bind specific chromatin-modifying enzymes: histone methyltransferases, deacetylases, and DNA methyltransferases. Inhibition of these enzymes passively affects gene regulation, can affect cellular processes outside of chromatin, and is therefore not ideal for direct epigenetic modulations. To achieve targeted epigenetic control of a broad range of repressed therapeutic genes, we constructed the first synthetic transcriptional activator, the Polycomb-based transcription factor (PcTF). Our previous work has shown that the PcTF fusion protein reads H3K27me3 DNA modification and activates a subset of H3K27me3-enriched genes. However, there remains a subset of those genes that are not upregulated by PcTF. Here, we present a new study to fully understand the mechanism of gene regulation through targeted interactions between chromatin reader-effectors and histones.

To fully elucidate the activation mechanism of PcTF we have developed stable MCF-7 cell lines that inducibly express full length PcTF or truncations of the full-length protein, to be utilized as controls. After induction, we harvested total mRNA at 10, 24, and 48 hours for RNA-seq. Bioinformatic analysis of the RNA-seq data stratified early and late-activated groups of target genes that specifically respond to full-length PcTF. This result shows that we can distinguish genes by their response to PcTF and establishes a foundation to investigate differences in the physical interaction of PcTF with specific promoters and enhancers, as well as differences in local chromatin structure. To this end, we are using CUT&RUN and a novel transposase mapping system (Calling Cards, R. Mitra lab) to track PcTF interactions and map chromatin features. Ultimately, our work will guide clinical use of synthetic reader-effectors to reprogram gene expression in cancer cells.

Timing the Cell: Towards a Synthetic Bio-Clock

Lai, G. - Presenter, Newcastle University

A bio-clock can be described as a built-in time keeping mechanism for biological systems. It would allow scientists to achieve temporal control of gene expression in living organisms, e.g. programming chasses to execute specific tasks at specific times. Implications for synthetic biology, both in industry and in the clinic, would be tremendous.

With a bio-clock, bacteria could be engineered to produce therapeutic molecules directly at the site of delivery, and at the desired time. In industry and bio-refineries, production pipelines could be streamlined and timed to match with more efficient workflows. We could even envisage microorganisms that are programmed to perform a complex series of tasks in a desired order just like computers run algorithms, thus paving the way to developments in bio-computing.

Despite many efforts, however, a bio-clock is still missing from the synthetic biology toolbox.

Here, we report how we have hijacked the natural circadian system of photosynthetic cyanobacteria to engineer a bio-clock in *E. coli*.

Towards Dynamic and Complex RNA Circuits: A Novel Approach to Address Retroactivity with Positive Feedback Insulation

Liu, B. - Presenter, Rice University

Chappell, J., Rice University

The creation of *de novo* designed RNA transcriptional regulatory systems with high dynamic range and orthogonality is allowing for the creation of RNA-based circuits that process signals solely at the transcriptional level. While much progress has been made, challenge remains to construct more complex and dynamic circuitry analogous to the ones achieved using protein-based regulators. One of the major challenges is high retroactivity seen in RNA regulatory systems. For example, in small RNA activating systems called STARs, a regulator RNA is irreversibly consumed for each regulatory event. To address this, we propose a novel approach to overcome retroactivity by implementing a positive-feedback insulation strategy that in essence compensates for the loss of regulatory RNAs. We present here a theoretical analysis of our strategy, called RNA compensator, and demonstrate that it can effectively mitigate the retroactivity not only in multi-state RNA circuits, e.g., self-activator and toggle switches, but also in dynamic RNA circuits like oscillators. In addition, we improve the connectivity between RNA modules by reshaping their input-output responses using tandem STAR and RNA sequestration. Finally, I will present our recent work creating broad-host RNA transcriptional regulators that use host-agnostic phage-derived RNA polymerases. Combining the works above, a toolset is developed to create modular, robust, and broad-host RNA circuits, providing a new platform for future synthetic RNA research.

Ultrasensitive Indicator Gas Reporting Enables Biosensing in Native Soil Conditions

Fulk, E. M. - Presenter, Northwestern University

Gao, X., Rice University

Redeker, K. R., University of York

Masiello, C. A., Rice University

Silberg, J. J., Rice University

Gas-producing enzymes can be used to noninvasively monitor the output of genetic circuits expressed in cells that are embedded in hard-to-image soils and sediments. For example, when expressed in environmental samples, methyl halide transferase (MHT) synthesizes volatile CH_3X that diffuses through the matrix and can be quantified in the sample headspace, enabling nondestructive monitoring of microbial behavior in opaque environments. To date, gas-reporting biosensors have been limited in use to autoclaved soils amended with halide salts and containing high titers of engineered microbes. However, an ideal biosensor for environmental applications could be added in small titers directly to a living (unautoclaved) soil without adding additional nutrients or halides. Here, we report on progress towards using gas-producing biosensors under more environmentally-reflective conditions. We apply gas concentration techniques to amply the signal of MHT-expressing microbes added to an autoclaved forest soil sample, and we explore the effects of halide concentration on CH_3X signal. We also report on progress towards detecting biosensor-produced gas signal above the background of a living soil. These studies advance the application of living sensors towards non-destructive studies of native environment samples and pave the way for understanding and controlling microbial behaviors in ecologically-important environments.

Using Cell-Free Protein Synthesis to Engineer Enzymes with Non-Canonical Amino Acids

Seki, K. - Presenter, Northwestern University

Zubi, Y., University of Chicago

Li, Y., University of Chicago

Roux, B., University of Chicago

Lewis, J., Indiana University Bloomington

Jewett, M., Northwestern University

Non-canonical amino acids (ncAAs) site-specifically decorate proteins with unique chemistries, enabling researchers to regulate protein activity, install unnatural metal-binding sites, and design new enzymes. However, due to inefficiencies in ncAA incorporation and screening workflows, proteins containing ncAAs remain difficult to engineer. In this work, we developed a protein engineering workflow comprised of molecular dynamics simulations, cell-free protein synthesis (CFPS), high-throughput liquid handling, and kinetic characterizations to study new functions encoded by ncAAs. As an example, we first studied how bipyridyl-alanine (Bpy), a metal chelating ncAA, could reversibly drive changes in enzyme activity. We show that, by combining CFPS with high-throughput liquid handling, we can efficiently synthesize dozens of enzyme mutants containing multiple Bpy residues and screen thousands of reaction conditions within days. In our model enzyme, prolyl oligopeptidase, we used this workflow to identify several phenotypes including both activation and inhibition by metals, study differential responses to a panel of divalent metal cations, and show that Bpy-metal complex formation is reversible by a competitive metal chelator. Finally, we generalize the workflow by using the same chemistry to control firefly luciferase activity. This work provides a rapid workflow to synthesize and screen for new protein functions encoded by ncAAs that will drive applications in protein engineering.

Using CRISPR/cas9 Editing to Explore the Spontaneous Adaptation of Ion Selectivity in a Bacterial Flagellar Motor

Baker, M. - Presenter, UNSW Sydney

Ridone, P., UNSW Sydney

Ishida, T., Hosei University

Sowa, Y., Hosei University

Motility provides a selective advantage to many bacterial species and is often achieved by rotation of flagella that propel the cell towards more favourable conditions. In most species, the rotation of the flagellum, driven by the Bacterial Flagellar Motor (BFM), is powered by H^+ or Na^+ ion transit through the torque-generating stator subunit of the motor complex. The ionic requirements for motility appear to have adapted to environmental changes throughout history but the molecular basis of this adaptation remains unknown. Here we used CRISPR engineering to replace the native *E. coli* H^+-powered stator with Na^+-powered stator genes and report the rapid and spontaneous reversion to H^+-power in a low sodium environment. We followed the evolution of the stators during their reversion to H^+-powered motility and used whole genome sequencing to identify both flagellar- and non-flagellar-associated genes involved in the cell's adaptation to new power sources. This work warns of the rapid reversion of CRISPR edits in rapidly evolving settings and highlights the utility of the flagellar stator system for studying the molecular mechanisms underlying adaptation, specifically demonstrating how environmental change can rapidly alter the function of an ion transporter.

Using Metabolic Network Modeling to Design an Escherichia coli strain with Enhanced Outer Membrane Vesicle Production

Ruiz-Moreno, H. A. - Presenter, Grupo de Disen?o de Productos y Procesos (GDPP)

Valderrama-Rincon, J. D., Universidad Antonio Nariño

González, A. F., Grupo de Disen?o de Productos y Procesos (GDPP)

Bacterial structures formed from the outer membrane and periplasmic components carry biomolecules to expel cellular material and interact with other cells. These outer membrane vesicles (OMVs) can encapsulate bioactive content, which gives OMVs a high potential as alternative drug delivery vehicles or novel vaccines. The biogenesis of OMVs has been associated with structural mechanisms such as interruptions in linkage between inner and outer membrane, pressure from accumulation of periplasmic material, and membrane curvatures induced by proteins.

Escherichia coli strains were engineered based on metabolic network modelling to find single gene deletion strains with enhanced OMVs production (Î"poxB, Î"sgbE, Î"gmhA, and Î"allD) through *in silico* gene knockout strain design applied to an already hypervesiculating strain (JC8031). Two deletions (Î"sgbE, Î"allD) have been linked to decreased amino acid precursor synthesis and decreased assimilation of nitrogenated compounds. The two other deleted genes (Î"poxB, Î"gmhA) are associated with lipopolysaccharide composition, lipid interaction, and intermembrane links. These deletions suggest that there is a relation between the metabolic activity and structural mechanisms that have been proposed as the cause of vesicle biogenesis.

These four mutants were obtained by genome editing using CRISPR-Cas9. Experiments on the recovery of OMVs from these strains showed enhanced OMV production in the four designed strains. Lipidomic analysis through LC-MS provided the lipid profile for E. coli OMVs. The lipid profile did not change significantly when comparing OMVs from the designed strains with the parent strain (JC8031).

The obtained results suggest that the vesicle production can be improved while the obtained vesicles are not altered in their composition, allowing further study for stability and integrity for use in therapeutic settings.

Using Newly-Identified Epigenetic Reader-Effector BAHCC1 to Engineer a H3K27me3-Binding Transcriptional Activator

Steppe, P. - Presenter, Georgia Institute of Technology

Haynes, K., Emory University

Epigenetic therapy has emerged as an approach to activate non-mutated tumor suppressor genes (TSGs) in cancer cells, which is an alternative to traditional chemotherapies that inhibit DNA replication. Currently, the design of epigenetic treatments for cancer is restricted to small molecules that bind the enzymes that generate biochemical marks (histone post-translational modifications and DNA methylation) within chromatin. Reader-effectors play a central role in translating these marks into gene expression states, but are less amenable to chemical inhibition. Our overall goal is to engineer reader-effectors that directly bind to modified histones in human cell nuclei for cancer epigenetics research and treatment. Our immediate goals are to identify binding modules with strong affinity and specificity for H3K27me3, and to use these modules to build reader-activators that can access TSGs in repressive chromatin and stimulate their expression. Previously we developed a synthetic reader-activator (SRA) protein with a binding module derived from the CBX8 protein. Doubling the number of CBX8 binding modules increases SRA activity at a H3K27me3-enriched reporter target. To further enhance SRA activity we set out to identify new functional variants of the CBX8-derived module. Here, we present an evolutionary analysis of metazoan CBX orthologs that identified 22 unique modifications that could be incorporated into our current SRA design to potentially increase binding affinity. We also considered a recently characterized reader-repressor, BAHCC1 (Fan et al. 2020, Nat. Genet.), which shows stronger H3K27me3-specific binding (Kd = 72.4 uM) than a CBX ortholog (Kd = 99.6 uM). We have isolated the H3K27me3-binding domain from BAHCC1 and incorporated it into an SRA. Functional characterization of SRAs that carry the new H3K27me3 modules is currently underway.

Using Qualitative Data for Quantitative Dynamics Estimation in Systems Biology

Ha, Y. - Presenter, Duke

Lu, J., Duke University

You, L., Department of Molecular Genetics and Microbiology

Wang, T., Duke University

Weiss, A., Duke University

Binary biological transitions are prevalent in microbial systems, ranging from the reversal of mobile genetic elementsâ€™ persistence to the change of complex speciesâ€™ interaction type. While these binary data are often generated, they are rarely used to assist quantitative dynamics information prediction. We investigate whether such binary qualitative data can be used to estimate quantitative data in these systems. We first demonstrate an approach to predict both qualitative and quantitative systems outcomes with mainly qualitative data via a â€œshape + structureâ€� model. In this approach, we use machine learning to predict the boundary of the transition with qualitative data. This boundary is then used to create contours, the shape, of estimated quantitative change. Some quantitative data points are used to provide additional information, the structure, of the real quantitative outcome. We use this structure to select the best machine learning model for both qualitative and quantitative predictions. To illustrate this approach, we predict the plasmid qualitative persistence and quantitative abundance information using mainly persistence data in a simple simulated community. We also show that plasmid dynamics in more complicated simulated communities and real experimental communities can be estimated in the same manner. Beyond this system, we demonstrate that a wide range of biology systems have similar property where qualitative information can be used to infer quantitative information. Our results indict the underused value of qualitative data in quantitative dynamics prediction in microbial communities.

Using Synthetic Biology for Determining Biological Controls on Marine Carbon Sequestration

Sridhar, S. - Presenter, Rice University

Ajo-Franklin, C., Lawrence Berkeley National Laboratory

Silberg, J. J., Rice University

Masiello, C. A., Rice University

A key flux in the marine carbon cycle is the transfer of photosynthesized organic matter from the surface ocean to the sea floor, which sequesters it in the deep ocean. This sequestration is mediated by complex biological processes that are regulated by bacterial quorum sensing. To understand the marine carbon flux, quantifying the bioavailable concentrations of autoinducer compounds that drive quorum sensing is imperative. Since current analytical methods fail to report these concentrations, we describe the construction and testing of bacterial biosensors that can persist in a marine matrix. These sensors detect acyl homoserine lactones and respond with a gas output, overcoming the limitation of fluorescence-based biosensors in marine matrices. With our biosensors, we can detect concentrations of acyl homoserine lactones in the nM range, which is biologically and environmentally significant. These new tools will enable measurements that provide quantitative insight into the biological controls of the marine carbon cycle and inform models of marine response to anthropogenic climate change.

Yeast Biosensors for Cyclic Peptide Drug Discovery By Directed Evolution

Scott, L. H. - Presenter, Astrazeneca

Gossing, M., Astrazeneca

Siewers, V., Chalmers University of Technology

David, F., Chalmers University of Technology

Wigglesworth, M. J., Astrazeneca

Davis, A. M., AstraZeneca

Small molecule drug discovery typically concentrates on diversity screening of constantly growing chemical libraries prior to lead optimisation using medicinal chemistry. However, failure to find hits against highly validated targets is forcing drug discovery efforts to explore new modalities such as proteins, peptides, and oligonucleotides. Discovery of these biology-based modalities can be enhanced by the power of natural selection via directed evolution, where genetic selections enable higher-throughput screening than possible with classical approaches. To harness this power of directed evolution, we developed a synthetic biology platform in the yeast *Saccharomyces cerevisiae* where genetic circuits can detect cyclic peptides generated *in vivo* that disrupt disease state protein-protein interactions important in human health. By using genetically encoded biosensors, each individual cell can act as a laboratory, thus creating a system where the analyses of billions of experiments can occur concurrently within a single culture. To do this, we developed a variation of the repressed transactivator yeast two-hybrid system to screen protein-protein interaction inhibitors. This selection incorporates a life or death pre-screen to purge negative hits. Building on this, we engineered a fluorescent readout dependent on the strength of protein interaction inhibition, enabling fine-tuned selection of potent inhibitors. As proof of concept, we show our system can detect the *in vivo* generated cyclic peptide *cyclo*-CLLFVY. This peptide is a known inhibitor of hypoxia inducible factor-1 (HIF-1) dimerisation, an important protein heterodimer regulator of hypoxia response in cancers. Future work will screen *in vitro* generated peptide libraries against other protein-protein interactions that are medically relevant, with the goal to move to the *in vivo* mutagenesis of peptide libraries for diversity creation. We argue this system will be generally applicable to any protein-protein interaction, ultimately aiming to deliver a versatile plug and play system for the continuous directed evolution and discovery of cyclic peptide drugs.

AUTHOR INDEX

AIChE
120 Wall Street, FL 23
New York, NY 10005-4020

ISBN 978-1-7138-4437-2